From plant genomics to plant biotechnology

Published by Woodhead Pub, 2013

Woodhead Publishing Series in Biomedicine

1. Practical leadership for biopharmaceutical executives
   J. Y. Chin
2. Outsourcing biopharma R&D to India
   P. R. Chowdhury
3. Matlab® in bioscience and biotechnology
   L. Burstein
4. Allergens and respiratory pollutants
   Edited by M. A. Williams
5. Concepts and techniques in genomics and proteomics
   N. Saraswathy and P. Ramalingam
6. An introduction to pharmaceutical sciences
   J. Roy
7. Patently innovative: How pharmaceutical firms use emerging patent law to extend monopolies on blockbuster drugs
   R. A. Bouchard
8. Therapeutic protein drug products: Practical approaches to formulation in the laboratory, manufacturing and the clinic
   Edited by B. K. Meyer
9. A biotech manager's handbook: A practical guide
   Edited by M. O'Neill and M. H. Hopkins
10. Clinical research in Asia: Opportunities and challenges
    U. Sahoo
11. Therapeutic antibody engineering: Current and future advances driving the strongest growth area in the pharma industry
    W. R. Strohl and L. M. Strohl
12. Commercialising the stem cell sciences
    O. Harvey
13. Biobanks: Patents or open science?
    A. De Robbio
14. Human papillomavirus infections: From the laboratory to clinical practice
    F. Cobo
15. Annotating new genes: From *in silico* screening to experimental validation
    S. Uchida
16. Open-source software in life science research: Practical solutions in the pharmaceutical industry and beyond
    Edited by L. Harland and M. Forster
17. Nanoparticulate drug delivery: A perspective on the transition from laboratory to market
    V. Patravale, P. Dandekar and R. Jain

Published by Woodhead Publishing Limited, 2013

18 Bacterial cellular metabolic systems: Metabolic regulation of a cell system with 13C-metabolic flux analysis
    K. Shimizu
19 Contract research and manufacturing services (CRAMS) in India: The business, legal, regulatory and tax environment
    M. Antani and G. Gokhale
20 Bioinformatics for biomedical science and clinical applications
    K-H. Liang
21 Deterministic versus stochastic modelling in biochemistry and systems biology
    P. Lecca, I. Laurenzi and F. Jordan
22 Protein folding *in silico*: Protein folding versus protein structure prediction
    I. Roterman
23 Computer-aided vaccine design
    T. J. Chuan and S. Ranganathan
24 An introduction to biotechnology
    W. T. Godbey
25 RNA interference: Therapeutic developments
    T. Novobrantseva, P. Ge and G. Hinkle
26 Patent litigation in the pharmaceutical and biotechnology industries
    G. Morgan
27 Clinical research in paediatric psychopharmacology: A practical guide
    P. Auby
28 The application of SPC in the pharmaceutical and biotechnology industries
    T. Cochrane
29 Ultrafiltration for bioprocessing
    H. Lutz
30 Therapeutic risk management of medicines
    A. K. Banerjee and S. Mayall
31 21st century quality management and good management practices: Value added compliance for the pharmaceutical and biotechnology industry
    S. Williams
32 Sterility, sterilisation and sterility assurance for pharmaceuticals
    T. Sandle
33 CAPA in the pharmaceutical and biotech industries: How to implement an effective nine step programme
    J. Rodriguez
34 Process validation for the production of biopharmaceuticals: Principles and best practice
    A. R. Newcombe and P. Thillaivinayagalingam
35 Clinical trial management: An overview
    U. Sahoo and D. Sawant
36 Impact of regulation on drug development
    H. Guenter Hennings
37 Lean biomanufacturing
    N. J. Smart
38 Marine enzymes for biocatalysis
    Edited by A. Trincone
39 Ocular transporters and receptors in the eye: Their role in drug delivery
    A. K. Mitra
40 Stem cell bioprocessing: For cellular therapy, diagnostics and drug development
    T. G. Fernandes, M. M. Diogo and J. M. S. Cabral

41 Oral delivery of insulin
   T. A. Sonia and Chandra P. Sharma
42 Fed-batch fermentation: A practical guide to scalable recombinant protein production in *Escherichia coli*
   G. G. Moulton and T. Vedvick
43 The funding of biopharmaceutical research and development
   D. R. Williams
44 Formulation tools for pharmaceutical development
   Edited by J. E. A. Diaz
45 Drug-biomembrane interaction studies: The application of calorimetric techniques
   Edited by R. Pignatello
46 Orphan drugs: Understanding the rare drugs market
   E. Hernberg-Ståhl
47 Nanoparticle-based approaches to targeting drugs for severe diseases
   J. L. Arias
48 Successful biopharmaceutical operations: Driving change
   C. Driscoll
49 Electroporation-based therapies for cancer: From basics to clinical applications
   Edited by R. Sundararajan
50 Transporters in drug discovery and development: Detailed concepts and best practice
   Y. Lai
51 The life-cycle of pharmaceuticals in the environment
   R. Braund and B. Peake
52 Computer-aided applications in pharmaceutical technology
   Edited by J. Petrović
53 From plant genomics to plant biotechnology
   Edited by P. Poltronieri, N. Burbulis and C. Fogher
54 Bioprocess engineering: An introductory engineering and life science approach
   K. G. Clarke
55 Quality assurance problem solving and training strategies for success in the pharmaceutical and life science industries
   G. Welty
56 TBC
57 Gene therapy: Potential applications of nanotechnology
   S. Nimesh
58 Controlled drug delivery: The role of self-assembling multi-task excipients
   M. Mateescu
59 *In silico* protein design
   C. M. Frenz
60 Bioinformatics for computer science: Foundations in modern biology
   K. Revett
61 Gene expression analysis in the RNA world
   J. Q. Clement
62 Computational methods for finding inferential bases in molecular genetics
   Q-N. Tran
63 NMR metabolomics in cancer research
   M. Čuperlović-Culf
64 Virtual worlds for medical education, training and care delivery
   K. Kahol

Woodhead Publishing Series in Biomedicine: Number 53

From plant genomics to plant biotechnology

EDITED BY
PALMIRO POLTRONIERI, NATALIJA
BURBULIS AND CORRADO FOGHER

Oxford   Cambridge   Philadelphia   New Delhi

Published by Woodhead Publishing Limited, 2013

Woodhead Publishing Limited, 80 High Street, Sawston, Cambridge, CB22 3HJ, UK
www.woodheadpublishing.com
www.woodheadpublishingonline.com

Woodhead Publishing, 1518 Walnut Street, Suite 1100, Philadelphia, PA 19102-3406, USA

Woodhead Publishing India Private Limited, G-2, Vardaan House, 7/28 Ansari Road, Daryaganj, New Delhi – 110002, India
www.woodheadpublishingindia.com

First published in 2012 by Woodhead Publishing Limited
ISBN: 978-1-907568-29-9 (print); ISBN 978-1-908818-47-8 (online)
Woodhead Publishing Series in Biomedicine ISSN 2050-0289 (print); ISSN 2050-0297 (online)

© The editors, contributors and the Publisher, 2013

The right of Palmiro Poltronieri, Natalija Burbulis and Corrado Fogher to be identified as authors of the editorial material in this Work has been asserted by them in accordance with sections 77 and 78 of the Copyright, Designs and Patents Act 1988.

British Library Cataloguing-in-Publication Data: A catalogue record for this book is available from the British Library.

Library of Congress Control Number: 2013932369

All rights reserved. No part of this publication may be reproduced, stored in or introduced into a retrieval system, or transmitted, in any form, or by any means (electronic, mechanical, photocopying, recording or otherwise) without the prior written permission of the Publishers. This publication may not be lent, resold, hired out or otherwise disposed of by way of trade in any form of binding or cover other than that in which it is published without the prior consent of the Publishers. Any person who does any unauthorised act in relation to this publication may be liable to criminal prosecution and civil claims for damages.

Permissions may be sought from the Publishers at the above address.

The use in this publication of trade names, trademarks, service marks, and similar terms, even if they are not identified as such, is not to be taken as an expression of opinion as to whether or not they are subject to proprietary rights. The Publishers are not associated with any product or vendor mentioned in this publication.
 The Publishers, editor(s) and contributors have attempted to trace the copyright holders of all material reproduced in this publication and apologise to any copyright holders if permission to publish in this form has not been obtained. If any copyright material has not been acknowledged, please write and let us know so we may rectify in any future reprint. Any screenshots in this publication are the copyright of the website owner(s), unless indicated otherwise.

Limit of Liability/Disclaimer of Warranty
The Publishers, editors and contributors make no representations or warranties with respect to the accuracy or completeness of the contents of this publication and specifically disclaim all warranties, including without limitation warranties of fitness of a particular purpose. No warranty may be created or extended by sales of promotional materials. The advice and strategies contained herein may not be suitable for every situation. This publication is sold with the understanding that the Publishers are not rendering legal, accounting or other professional services. If professional assistance is required, the services of a competent professional person should be sought. No responsibility is assumed by the Publishers, editors or contributors for any loss of profit or any other commercial damages, injury and/or damage to persons or property as a matter of products' liability, negligence or otherwise, or from any use or operation of any methods, products, instructions or ideas contained in the material herein. The fact that an organisation or website is referred to in this publication as a citation and/or potential source of further information does not mean that the Publishers nor the editors and contributors endorse the information the organisation or website may provide or recommendations it may make. Further, readers should be aware that internet websites listed in this work may have changed or disappeared between when this publication was written and when it is read. Because of rapid advances in medical sciences, in particular, independent verification of diagnoses and drug dosages should be made.

Typeset by RefineCatch Limited, Bungay, Suffolk
Printed in the UK and USA

Contents

List of figures	xiii
List of tables	xv
List of abbreviations	xvii
About the contributors	xxi

	Introduction		1
1	**From plant genomics to -omics technologies**		3
	Palmiro Poltronieri, Institute of Sciences of Food Productions, ISPA-CNR, Italy		
	1.1	SuperSAGE	5
	1.2	CAGE – cap analysis of gene expression	6
	1.3	-Omics and new advances in plant functional genomics	7
	1.4	Bibliography	10
2	**Plant microRNAs**		15
	Moreno Colaiacovo and Primetta Faccioli, Consiglio per la Ricerca e la sperimentazione in Agricoltura, Centro di ricerca per la genomica e la postgenomica animale e vegetale, Fiorenzuola d'Arda, Italy		
	2.1	Introduction	16
	2.2	Transcription of miRNA genes	16
	2.3	MicroRNA processing	17
	2.4	Modes of action	18
	2.5	Evolution of miRNA genes	19
	2.6	Differences from animal miRNAs	20
	2.7	MiRNA functions	20
	2.8	The potential roles of microRNAs in crop improvement	25
	2.9	Bibliography	26

| 3 | Epigenetic control by plant Polycomb proteins: new perspectives and emerging roles in stress response | 31 |

Filomena de Lucia, Institut Pasteur, France, and Valérie Gaudin, Institut Jean-Pierre Bourgin, France

3.1	Introduction	32
3.2	Conserved multi-protein complexes with histone post-translational modifying activities	32
3.3	Polycomb functions in plant development	37
3.4	Non-coding RNAs as regulatory cofactors of Polycomb complexes	38
3.5	Emerging roles of PcG and ncRNAs in responses to environmental stress	40
3.6	PcG proteins functions in three-dimensional nuclear organization	41
3.7	Perspectives: a role of Polycomb in abiotic and biotic stress response	42
3.8	References	43

| 4 | Metabolite profiling for plant research | 49 |

Nalini Desai and Danny Alexander, Metabolon, Inc., USA

4.1	Introduction	49
4.2	Methodological approach	50
4.3	Metabolomic platform	52
4.4	Metabolomics in plant science	54
4.5	The future role of metabolomics in crop improvement	55
4.6	Conclusion	59
4.7	References	60

| 5 | The uniqueness of conifers | 67 |

Carmen Díaz-Sala, Department of Plant Biology, University of Alcalá, Spain, José Antonio Cabezas, National Research Institute for Agricultural and Food Technology (INIA), Spain, Brígida Fernández de Simón, National Research Institute for Agricultural and Food Technology (INIA), Spain, Dolores Abarca, Department of Plant Biology, University of Alcalá, Spain, M. Ángeles Guevara, National Research Institute for Agricultural and Food Technology (INIA), Spain, Mixed Unit of Forest Genomics and Ecophysiology, INIA/UPM, Spain, Marina de Miguel, National Research Institute for

Agricultural and Food Technology (INIA), Spain, Estrella Cadahía, National Research Institute for Agricultural and Food Technology (INIA), Spain, Ismael Aranda, National Research Institute for Agricultural and Food Technology (INIA), Spain, and María-Teresa Cervera, National Research Institute for Agricultural and Food Technology (INIA), Spain, Mixed Unit of Forest Genomics and Ecophysiology, INIA/UPM, Spain

5.1	Introduction	68
5.2	Functional differentiation	69
5.3	Genome structure and composition	72
5.4	Genome function	76
5.5	Chemical divergence	82
5.6	Meeting the challenge: the system biology approach to unraveling the conifer genome	85
5.7	Acknowledgements	86
5.8	References	86

6 **Cryptochrome genes modulate global transcriptome of tomato**   97
Loredana Lopez and Gaetano Perrotta, ENEA, Trisaia Research Centre, Italy

6.1	Introduction	97
6.2	Cryptochrome functions	99
6.3	Role of cryptochromes in mediating light-regulated gene expression in plants	100
6.4	Cryptochromes influence the diurnal global transcription profiles in tomato	105
6.5	References	110

7 **Genomics of grapevine: from genomics research on model plants to crops and from science to grapevine breeding**   119
Fatemeh Maghuly, BOKU VIBT, Austria, Giorgio Gambino, Plant Virology Institute, National Research Council (IVV-CNR), Italy, Tamás Deák, Corvinus University of Budapest, Hungary, and Margit Laimer, BOKU VIBT, Austria

7.1	Use of genetic and molecular markers for studies of genetic diversity and genome selection in grapevine	120
7.2	Grapevine breeding	123
7.3	Transgene silencing	128

7.4	Identification and characterization of transgene insertion loci	131
7.5	Integration of vector backbone	133
7.6	Stability of inserted transgenes	135
7.7	Conclusions	137
7.8	Acknowledgement	137
7.9	References	138

8 Grapevine genomics and phenotypic diversity of bud sports, varieties and wild relatives — **149**
Gabriele Di Gaspero and Raffaele Testolin, Dipartimento di Scienze Agrarie e Ambientali, University of Udine, Italy and Institute of Applied Genomics / Istituto di Genomica Applicata, Parco Scientifico e Tecnologico Luigi Danieli, Italy

8.1	Introduction	150
8.2	Origin of *Vitis vinifera*, domestication, and early selection for fruit characters	151
8.3	Sources of phenotypic variation in present-day grapevines	153
8.4	Genomic tools in the genome sequencing era	153
8.5	Current activities in grapevine genome analysis	154
8.6	Bud organogenesis, somatic mutations, and DNA typing of somatic chimeras	155
8.7	Phenotypically divergent clones and the underlying DNA variation	156
8.8	Transposon insertion-site profiling using NGS	158
8.9	Large structural variation using NGS	158
8.10	Copy number variation, gene redundancy, and subtle specialisation in secondary metabolism	159
8.11	Conclusions	160
8.12	References	160

9 Peach ripening transcriptomics unveils new and unexpected targets for the improvement of drupe quality — **165**
Nicola Busatto, Abdur Md Rahim and Livio Trainotti, University of Padova, Italy

9.1	Introduction	166
9.2	The fruit	167

	9.3	Peach development and ripening	169
	9.4	Microarray Transcript Profiling in peach	170
	9.5	New players in the control of peach ripening	174
	9.6	Conclusions	178
	9.7	Acknowledgements	179
	9.8	References	179

10 Application of doubled haploid technology in breeding of *Brassica napus* — 183
Natalija Burbulis, Aleksandras Stulginskis University, Lithuania, and Laima S. Kott, University of Guelph, Canada

10.1	Introduction	184
10.2	Technique of isolated microspore culture	184
10.3	Doubled haploid method in breeding of *Brassica napus*	185
10.4	In vitro mutagenesis	186
10.5	Utilization of double haploidy in selection for resistance	188
10.6	Selection for modified seed oil composition	191
10.7	Selection for improved seed meal	193
10.8	Selection for cold tolerance	196
10.9	Concluding remarks	197
10.10	References	197

11 Plant biodiversity and biotechnology — 205
Naglaa A. Ashry, Field Crops Research Institute, ARC, Egypt

11.1	Biodiversity	205
11.2	Biotechnology	206
11.3	Heat stress tolerance in cereals	207
11.4	Modern approaches in cereals for yield and food security under temperature stress	210
11.5	Future perspectives	215
11.6	Acknowledgment	215
11.7	References	216

12 Natural resveratrol bioproduction — 223
Angelo Santino, Marco Taurino, Ilaria Ingrosso and Giovanna Giovinazzo, Institute of Sciences of Food Productions, CNR-ISPA, Italy

12.1	Stilbenes and resveratrol	223

12.2 Health benefits of resveratrol	225
12.3 *trans*-resveratrol production through plant cell cultures	226
12.4 Introducing new pathway branches to crop plants: *trans*-resveratrol synthesis in tomato fruits	228
12.5 Concluding remarks	230
12.6 References	231
Index	235

List of figures

1.1	SuperSAGE protocol scheme: cDNA cleavage, formation of DNA tags and ligation into ditags	5
1.2	Approaches to applications of systems biology and -omics technologies in plant crops improvement	7
2.1	miRNA biogenesis in plant cells	17
2.2	Main plant miRNAs involved in stress response and development	21
3.1	Core components of the PRC complexes in Drosophila and *A. thaliana*	33
3.2	The vernalization process and the PRC2 complexes	35
3.3	Models of ncRNA acting *in cis* (A) and *trans* (B) to recruit PRC2 complexes	39
5.1	Key biological differences between conifers and angiosperms	70
5.2	The distribution of genome sizes in angiosperms and conifers	72
7.1	Expression cassettes of plant transformation vectors carrying different sequences of the GFLV CP gene	127
9.1	Expression profiles of ethylene-related genes during fruit development and ripening	172
9.2	Chromosomal location of ethylene-related genes on the peach physical map	173
12.1	Resveratrol biosynthesis pathway	224

List of tables

12.1	Bioproduction of resveratrol by plant cell cultures	227
12.2	Bioproduction of resveratrol by plant metabolic engineering	229

Abbreviations

3C	chromosome conformation capture
A2C	azetidine-2-carboxylate
ABA	abscisic acid
AGO	ARGONAUTE
amiRNA	artificial microRNA
ANOVA	analysis of variance
antagomiR	a sequence that complexes and sequesters a microRNA
ARE	auxin responsive elements
ArMV	Arabis mosaic virus
BAC	bacterial artificial chromosome
CAGE	cap analysis of gene expression
CBC	cap-binding complex
CBD	cyclobutane-pyrimidine-dimer
CD	β-cyclodextrin
cDNA	copy DNA
ceRNA	competing endogenous RNAs
CHS	chalcone synthase
CI	chilling injury
CIMMYT	International Maize and Wheat Improvement Center
CML	CIMMYT Maize Line
COS	Conserved Ortholog Sets
COST	European Cooperation in Science and Technology
COX	cyclooxygenase
cpDNA	conifer chloroplast DNA
CP-MR	coat protein-mediated resistance
CRY	cryptochrome
CSD2	Cu-Zn superoxide dismutase 2
CSPW	cabbage seedpod weevil
DCL	DICER-like RNAse
DH	double haploid
DI	Incidence of Disease
DIR	dirigent proteins

diTPS	diterpene synthase
dN/dS	non-synonymous distance / synonymous distance
DP	3,4-dehydro-D,L-proline
dsRNA	double-stranded RNA
ELISA	enzyme-linked immunosorbent assay
EMS	ethyl methanesulfonate
ENU	ethylnitrosourea
eQTL	expression QTL
EST	expressed sequence tag
FAO	Food and Agriculture Organization
FPA	p-fluoro-D,L-phenylalanine
GA	gibberellic acid
GC	gas chromatography
GFLV, GFV	Grapevine fanleaf virus
GI	Gigantea
GMP	genetically modified plants
GUS	beta-glucuronidase
GWAS	genome-wide association studies
HMT	histone methyltransferase
HP	hydroxyproline
HPLC	high performance liquid chromatography
hpRNA	hairpin RNA
HSF	heat stress transcription factor
HSP	heat shock protein
IAA	auxin
IAEA	International Atomic Energy Agency
INIA	National Research Institute for Agricultural and Food Technology
INR	transcription initiator
IPK	Institut für Pflanzen Genetik und Kultur Pflanzen Forschung
IPM	integrated pest management
ISPA-CNR	The Institute of Food Production of the National Research Council
IR	interspersed repeat
JA	jasmonate
LB	left border
LC	liquid chromatography
LD	linkage disequilibrium
LecRK	lectin receptor kinase
lncRNA	long interspersed non-coding RNAs

LRR-RLK	leucine-rich-repeat
LTR	long terminal repeat
MAS	marker assisted selection
1-MCP	1-methylcyclopropene
miRNA	microRNA
MJ	methyl jasmonate
MP	movement protein
MS	mass spectrometry
NAA	1-naphthalene acetic acid
NCED	9-cis-epoxycarotenoid dioxygenase
ncRNA	non-coding RNA
NGS	next-generation sequencing
NLN	Nitsch Liquid Nutrient
NMR	nuclear magnetic resonance
nt	nucleotide
NUE	nitrogen use efficiency
ORF	open reading frame
PcG	Polycomb group
PCR	polymerase chain reaction
PDR	pathogen derived resistance
PHD	plant homeodomain
PHY	phytochrome
P_i	orthophosphate
PRE	Polycomb response element
PSI	Photosystem I
PSK	phytosulfokine
PSR	P-starvation-responsive
PTGS	post-transcriptional gene silencing
QPM	Quality Protein Maize
QRT-PCR	quantitative real-time polymerase chain reaction
QTL	Quantitative Trait Loci
RAPD	random amplified polymorphic DNA marker
RB	right border
RdDM	RNA-directed DNA methylation
RdRp	RNA-dependent RNA polymerase
RFLP	restriction fragment length polymorphism marker
RGF	root growth factor
RIP	RNA immunoprecipitation
RISC	RNA-induced silencing complex
RITS	RNA-induced transcriptional gene silencing
RLK	receptor like kinase

ROS	reactive oxygen species
RRM	reiterated reproductive meristem
Rubisco	ribulose 1,5-bisphosphate carboxylase/oxygenase
SA	salicylic acid
SAGE	serial analysis of gene expression
SENESCO	a company name
sHSP	small heat-shock proteins
SI	self-incompatibility
SIGA	Italian Society of Agriculture Genetics
siRNA	short (or small) interfering RNA
SNP	single nucleotide polymorphism
sRNA	small RNA
SSR	simple sequence repeat
STS	stilbene synthase
SYS	systemin
tasiRNA	*trans*-acting siRNA
T-DNA	transferred DNA
TE	transposable elements, tracheary element
TGS	transcriptional gene silencing
Ti	tumor-inducing
TILLING	Targeting Induced Local Lesions In Genomes
TMV	Tobacco mosaic virus
UHPLC	ultra high performance liquid chromatography
USDA	US Department of Agriculture
UTR	untranslated region
VBS	Vector backbone sequences
VMC	*Vitis* Microsatellite Consortium

About the contributors

Dr Dolores Abarca is an Associate Professor in the Plant Biology Department, University of Alcalá, Spain. She has spent more than 25 years studying the molecular basis of biological processes. She initially trained as a pharmacist, before taking an MSc and a DPhil in Molecular Biology. She worked as a post-doctoral researcher at Harvard University and at the National Institute for Agronomic Research (Madrid). Her research interests are focused on the regulatory pathways that connect plant development and responses to the environment. She is currently involved in research projects aimed at adapting biotechnology techniques developed in model plant species to unravel the regulation of basic functions in conifers. Her research has been funded by national and international public agencies, including the European Commission. In addition, she has been teaching Plant Biotechnology for the last ten years.

Dr Danny Alexander has over 30 years' experience in plant molecular biology and biochemistry. Before joining Metabolon, he held positions at the ARCO Plant Cell Research Institute and Calgene, Inc., working on cloning of agriculturally important genes, and at Ciba-Geigy/Novartis/Syngenta, and BASF, doing research on plant disease resistance, maize molecular genetics, and genomics. Danny has been at Metabolon since its inception in 2002, helping develop the metabolomics platform and software, as well as providing biological interpretation of metabolomic data to clients in a wide range of medical and plant-related disciplines. He currently serves as Senior Project Manager at Metabolon.

Dr Ismael Aranda is a PhD specialist in the ecophysiology of forest tree species, working at the Forest Research Centre of INIA (Madrid). He is an experienced researcher who has dealt, in the last 20 years, with the functional response and phenotypic plasticity of plants to different environmental factors. He has been especially concerned with the impact of drought at different ontogeny states, from the seedling to the mature tree. His research interest also spans the study of the intra-specific adaptive variation to drought within different forest tree species. Ismael is the co-author and author of different book chapters and over 50 peer-reviewed papers published in the top scientific journals of different fields from Forestry, Ecology and Plant Biology, including *Tree Physiology* (he has been a member of the official editorial board since 2008), *Tree Genetics and Genomes*, *Environmental and Experimental Botany*, and *Molecular Ecology*. He has been the supervisor of several PhD theses, and currently leads education and training of several PhD scholarship holders. He has participated (leading some of them as PI) in some regional, national and international projects funded by the Spanish Ministry of Economy and Competitiveness and from the FP7 UE.

Dr Naglaa Abdel-monem Mahmoud Ashry is Professor of Genetics, and Head of the Cell Research Section, Field Crops Research Institute, Agricultural Research Center, Egypt. She obtained her PhD from the Genetics Department, College of Agriculture, Ain Shams University, Cairo, Egypt, in 1998. Her thesis was entitled 'Genetic Studies on Some Flax Properties under Some Environmental Stresses'. Her experience includes being PI of a research project entitled 'Isolation of powdery mildew resistance gene(s) in flax', funded by EU grant S2/J8/20 (EXTERNAL ACTIONS OF THE EUROPEAN COMMUNITY), which ended in 2010. She participated in the preparation of and negotiations for an ICI project in cooperation with Helsinki University and the Finnish Agricultural Research Center (MTT); the project is financed by the Finnish Ministry of Foreign Affairs and is due to start in January 2013. She participated in the preparation of an FP7-KBBE project 'Biowaste 4SP,' in a consortium of the EU countries, Egypt, Morocco, Tunisia, and Tanzania. The project was approved and is being implemented.

Dr Natalija Burbulis is Head of the Agrobiotechnology laboratory and Professor at the Crop Science and Animal Husbandry Department of the Aleksandras Stulginskis University, Lithuania. She holds a PhD in Agricultural Science obtained from the Lithuanian University of Agriculture, and for ten years has performed research on plant biotechnology, physiology and biochemistry. Her current studies are on in vitro selection of oilseed crops (rapeseed and linseed) genotypes with important agronomic traits, including disease resistance, cold tolerance and oil quality improvements. She was awarded the prestigious NATO Science Fellowship, administered by the NSERC (Natural Sciences and Engineering Research Council of Canada), and was a Fellow (2003–2004) in the Department of Plant Agriculture of the University of Guelph, Canada. She is a member of the

European Association for Research on Plant Breeding. She is the author or co-author of more than 80 peer-reviewed papers.

Dr Nicola Busatto was born in Padova, Italy, in 1982. At the University of Padova, he completed a Bachelor's Degree in Molecular Biology in 2006 and a Master's Degree in Molecular Biology in 2008. He obtained his PhD in Crop Science (Agrobiotechnology) in 2012, working in the Plant Functional Genomics laboratory at the Department of Biology of the University of Padua, under the supervision of Dr Livio Trainotti. He carried out the functional characterization of a peach RGF-like peptide hormone, deepening understanding of the underlying aspects of the auxin–ethylene cross-talk during fruit ripening.

Dr José Antonio Cabezas is a researcher focused on the genetic and genomic analysis of woody plant species, including both cultivated, such as grapevine or olive, and forest, such as different pine or eucalyptus, species. His area of expertise comprises the use of advanced genomic tools to better understand their genetic diversity and the evolution of their genomes, as well as identification of the genetic determinants of relevant quality and productive traits. An important part of his work has been targeted to the development and transference of knowledge to breeders, for example, to allow the early selection of the individuals carrying the most favourable allelic combinations for a specific trait via marker-assisted selection. He has developed his professional career in public research institutions, mainly INIA, IMIDRA and CNB, and has a PhD in Biology from the Universidad Autónoma de Madrid (Spain).

Dr Estrella Cadahía is a Scientist at Forest Research Centre, in the National Research Institute for Agricultural and Food Technology, Madrid, Spain. She has over 30 years experience in research on the chemical composition of vegetables in relation to their functionality and food. She began her career at Vegetable Protection as a chemist, before taking a PhD thesis in Forestry Chemistry (Universidad Complutense, Madrid). Her scientific experience spans: the quality of agro-forestry products, the chemistry of wood and cork in relation to their geographical origin, industrial processing and its use in oenology; chemical and sensorial characteristics of wine in relation to wood used during its processing; and on the other hand, interaction between the chemical composition of forestry species (leaf, wood and bark or resin, needle and wood in conifers), and environmental, physiological and genetic factors, including the metabolomics approaches. She has coordinated and collaborated on numerous research projects, financed by the Spanish Government, and on various technological research projects of companies in related sectors. She has published numerous articles in the top scientific journals in her field, and she has acted as review member for the National Evaluation and Foresight Agency (ANEP) and other private and public agencies, as well as for several top scientific journals.

Dr María Teresa Cervera is a senior researcher at INIA (Spain), specializing in Forest Genomics. The primary research aim of her team is to unravel the molecular control of forest tree adaptive responses, combining functional and structural genomics (i.e. gene discovery, SNP genetic mapping, QTL analysis and Candidate Gene characterization by gene expression and nucleotide variability), epigenetic analysis and comparative genomics. She has worked in several institutions worldwide, such as CBM-CSIC and CNB-CSIC (Spain), RUG (Belgium), SRI and NCSU (USA), and NIFTS (Japan). Her team has also developed active collaborations with private companies as well as with national and regional administrations, focused on the development and application of genomic resources for molecular breeding in woody plant species. She has participated in numerous international and national projects, leading several initiatives, including projects on conifer genome sequencing. She has a BSc and a PhD in Molecular Biology, and has authored over 80 SCI papers and book chapters.

Dr Moreno Colaiacovo is PhD fellow at the University of Turin (Italy). Since 2009, he has been working as a computational biologist at the 'Consiglio per la Ricerca e la Sperimentazione in Agricoltura (CRA) – Centro di ricerca per la genomica e la post-genomica animale e vegetale' located in the North of Italy. Moreno has a Bachelor's degree in Biotechnology from the University of Insubria and a Master's degree in Bioinformatics from the University of Milan-Bicocca. During his work as a computational biologist he has studied plant microRNAs in several plant species with different bioinformatic approaches. He has applied computational pipelines to identify microRNAs both in genomes and in transcriptomes related to species of agronomical interest, and has studied regulatory circuits involving transcription factors and microRNAs.

Dr Tamás Deák, PhD, is Assistant Professor in the Department of Viticulture of the Corvinus University of Budapest. He has specialized in plant breeding and molecular genetics with a particular interest in bioinformatics. He has been teaching genetics, breeding and biological resources of viticulture for over ten years. His research experiences cover structural and functional genomics of grapevines; development and application of genetic and genomic markers for grape breeding; molecular functions of plant–pathogen interactions; and winter hardiness of grapes. He is currently analysing the functional roles of host genes in grape–Agrobacterium interactions.

Dr Filomena de Lucia is an experienced scientist who has worked in the chromatin and epigenetics field since the beginning of her career at the University of Naples, where she obtained her PhD. She then left Italy to work in several prestigious European institutes using different model systems. She spent five years at the John Innes Centre in Norwich (UK), working on epigenetic phenomena in *Arabidopsis thaliana*, with a focus on Polycomb proteins and the link with non-coding RNAs and nuclear organization. While there, she organized the UK meeting 'Plant chromatin and nuclear organization', held at the John Innes Centre and sponsored by Millipore. She is currently working at the Pasteur Institute in Paris, applying her knowledge in the field of pathogens. She holds a PI diploma and a professor qualification.

Dr Marina de Miguel is a PhD student at INIA-CIFOR – National Institute of Agricultural Research-Forest Research Centre in Madrid, Spain. Her PhD project deals with the genetic control of functional characters in response to drought in a Mediterranean pine species, *Pinus pinaster*. Her main interests are the construction of genetic linkage maps, identification of QTLs, and the functional and biochemical response to drought in forest species, especially in conifers. During her PhD she spent two short periods at INRA-Nancy and INRA-Bordeaux (France) to work on the construction of a consensus genetic linkage map of *Pinus pinaster*. She has worked in the identification of orthologous markers between *Pinus pinaster* and *Picea glauca* during a short stay in CFL-RNC (Québec, Canada). She has a Bachelor's degree in Biological Science from the Universidad Complutense de Madrid and a Master's degree in Environmental Science and Technology from the Universidad Rey Juan Carlos.

Dr Nalini Desai is currently a Senior Study Director at Metabolon, involved in interpreting the biological relevance of metabolomic data for clients in the field of agriculture and plant sciences. Prior to joining Metabolon, she held various positions in the field of agricultural biotechnology, first at Ciba/Novartis/Syngenta, where she played a key role in the development of the first transgenic maize product, and most recently as a project leader at Athenix Corp. and Bayer Crop Science. In that role, she successfully led her team in the discovery and development of novel insect control traits. She continues to be interested in the application of technology for engineering crops with improved field performance.

Dr Carmen Díaz-Sala is Associate Professor at the Department of Plant Biology, University of Alcalá, Madrid, and an expert in physiological and molecular regulation of tree development. She leads the AgroForestry Biotechnology group at the University of Alcalá. Her research interest is focused on functional genomics and the molecular and physiological regulation of the developmental processes underlying age and maturation-associated traits involved in forest productivity. She has developed a complementary experimental approach including physiological, biochemical, cellular, molecular and genomics analysis to study the effect of maturation on the propagation capacity and performance of forest reproductive materials. She is the Principal Investigator of several projects and initiatives.

Dr Gabriele Di Gaspero is Associate Professor of Grapevine Genetics and Breeding at the University of Udine, Italy. He holds a PhD in Plant Biotechnology, and in the early 2000s he took over the management of a grapevine breeding program for introgressing disease-resistant genes into high quality wine grapes. Initially active in the development and exploitation of molecular markers to assist breeding, since the completion of the grape genome sequencing he has progressively become involved in the analysis of genetic diversity and genome evolution in wine grapes.

Dr Primetta Faccioli is a researcher at the 'Consiglio per la Ricerca e la Sperimentazione in Agricoltura (CRA)-Centro di ricerca per la genomica e la post-genomica animale e vegetale', located in the North of Italy. She has a Master's degree in Biology from the University of Pavia, a diploma of Specialization in Applied Genetics from the University of Milan and a postgraduate Master's degree in Bioinformatics from the University of Turin. She has spent about 20 years studying plant genetics and researching how to apply molecular biology to plant breeding. In 2004 she moved from the laboratory to the computer room to apply computational biology techniques to her research activity, which is mainly related to the study of complex regulatory circuits involving microRNAs and transcription factors in plants of agronomical interest.

About the contributors

Dr Brígida Fernández de Simón is a Senior Researcher at the Forest Research Centre, at the National Research Institute for Agricultural and Food Technology (INIA), Madrid, Spain. She has spent more than 25 years studying and researching the relations between the chemical composition of vegetables and food and their functionality. She initially trained as a pharmacist, before taking a PhD in Food Science and Technology, at the Scientific Research Superior Council (CSIC) and the Complutense University (both in Madrid). Her research interests cover the quality of wines and grapes, the chemistry of wood and cork in relation to their geographical origin, their industrial processing and their use in oenology, as well as the interaction between the chemical composition of trees (resin, needle and wood in conifers, leaf, wood and bark in other species) and environmental, physiological and genetic factors, including metabolomics approaches. She has coordinated and collaborated on numerous research projects, funded by the Spanish Government, as well working for as various companies in related sectors. She has published in the top scientific journals in her field, and has consulted for the USDA Viticulture Consortium and the American Vineyard Foundation, as well as providing advice to wineries, coopers, other companies, and institutional departments in Spain.

Dr Giorgio Gambino is a researcher at the Plant Virology Institute, National Research Council (IVV-CNR), Grugliasco Unit, Italy. He received his PhD from the University of Turin in 2004, with his dissertation on the genetic transformation of grapevine. He continued his research on the genetic transformation and molecular characterization of transgenic plants of grapevine, tobacco and Arabidopsis for virus resistance and for functional genomics studies. In addition, his main research activities are: induction of somatic embryogenesis in grapevine, and studies of the molecular bases of embryogenesis (WOX genes); RNA silencing and small RNA characterization; detection and eradication of viruses, viroids and phytoplasma in grapevine; and molecular interactions between viruses and grapevine.

Dr Valérie Gaudin is a senior scientist in Plant Molecular Biology working at the Jean-Pierre Bourgin Institute (INRA Versailles). She is particularly interested in epigenetics and chromatin gene regulation. Her group is currently studying Polycomb repression mechanisms in the plant model *Arabidopsis thaliana*. She has published in top scientific journals in this field, and authored a number of research papers focusing on epigenetics and on methodologies to study protein–DNA interactions.

Dr Giovanna Giovinazzo holds a degree in Biological Science, in plant genetics. She was a research fellow at the Institute of Plant Biosynthesis (Milan), funded by the Strategic Project 'Agrotechnology' of the National Research Council of Italy, working on 'Studies of agronomically useful genes in crops'. She was a researcher at the Institute of Research on Agro-Food Biotechnology (IRBA), Lecce, focusing on 'Improvement of nutritional quality of crops and in vitro production of plant metabolites through innovative biotechnology' and project leader for 'Improvement of food quality of tomato plants modulating the flavonoids pathway'. She is a Senior Researcher at ISPA-CNR, in the field of plant antioxidant metabolism related to nutrition and human health, metabolic engineering for the improvement of the nutritional value of staple plants (tomato fruits), evaluation of the functional properties of specific components/ingredients of traditional foods (polyphenols in grape skin and wine), and 'Bioactive compounds in traditional and innovative food system (polyphenols)'. She is a member of the Italian Association of Agricultural Genetics (SIGA), Group Polyphenols, Working Group 'Plant as Bioreactors'.

Dr M. Ángeles Guevara is a researcher at the Forest Research Centre (CIFOR), in the National Research Institute for Agricultural and Food Technology (INIA), Madrid, Spain. Her research experience and interest deal, fundamentally, with the development of molecular markers, epigenetic studies, the characterization of candidate genes, studies in diversity, genetic maps and transcriptomics. Currently, she is focusing on functional genomics for adaptation and the epigenetic control of the response to drought in forest tree species. She has developed her professional career in public research institutions such as the Universidad Politécnica de Madrid, CIB (CSIC) and INIA. She has participated in many regional, national and international projects funded by regional public institutions, the Spanish Ministry of Economy and Competitiveness and the UE (FP7), and authored several papers. She has a BSc and a PhD in Biology from the Universidad Complutense de Madrid.

Dr Ilaria Ingrosso holds a degree in Biology, with her thesis in Biochemistry 'Analysis of Complex III oligomeric status of Saccharomyces cerevisiae mitochondrial respiratory chain'. She was a research fellow working on improving the nutritional and organoleptic properties of gluten-free foods, carried out at the Institute of Food Production of the National Research Council (ISPA-CNR), and on the 'Improvement of food quality and health promoting properties of tomato tissue, wine and grapevine'.

Currently, she holds a fellowship on in vitro production of trans-resveratrol by grape cell cultures at ISPA-CNR.

Dr Laima S. Kott has been involved in doubled haploid research since 1980, first working with barley, then with canola and most recently with wheat and maize. Over the past six years she has developed 12 canola cultivars for Canada and global commercialization. She developed a number of specialty canola germplasms derived from crosses from weedy relatives through introgression of traits for insect resistance (e.g. cabbage seedpod weevil; root maggot). Furthermore, she has produced novel protocols for in vitro selection of Sclerotinia resistance, frost tolerance in spring canola, and low saturated fats in canola as well as Fusarium resistance in wheat. In 1997, she released a patent on hybrid production using self-incompatibility alleles in canola. Currently she is developing mint clones with super-high antioxidant levels destined for the nutraceutical and medical industries.

Dr Prof. Margit Laimer, Head of the Plant Biotechnology Unit (PBU), University of Natural Resources and Life Sciences (BOKU), Vienna, Austria, since 1987, is an expert in plant biotechnology. She holds a PhD in Botany and Zoology (Univ. Vienna, 1985) and two habilitations in two related fields, Plant Biotechnology (1991, BOKU, Vienna) and Plant Virology (1993, Univ. Lisbon). She has acquired extensive competence in plant engineering since 1985 when confronted with the challenge of the production of healthy food employing the development of methods for rapid detection of plant pathogens, and the reduction of the use of biocidal chemicals via the employment of strategies which improve the natural resistance of plants. Since 1990, strategically important viruses have been included in her research portfolio, as well as plant tissue cultivation and vaccine production in plants. She has published a series of highly relevant scientific manuscripts in renowned plant and horticultural journals (more than 160 papers), and actively contributed to international conferences. She has published a book and many book chapters, and she serves on the editorial board of many journals. She has also invested considerable time in communicating issues connected with GMO technology to the general public in Austria and in Europe.

Dr Loredana Lopez has a PhD in Plant Biology. She has more than ten years of experience in the plant molecular biology and biotechnology fields. Her research interests span genomic and proteomic investigation in several plant species, as well as ultra-massive genome and transcriptome sequencing of cultivated plants. She has published in top scientific journals in this field, including *Nature*, and has authored a number of research papers focusing on tomato photoreceptors. At present she has a post-doctoral position at ENEA Research Centre, Italy.

Assoc. Prof. Dr Fatemeh Maghuly, Deputy Head of the Plant Biotechnology Unit (PBU), University of Natural Resources and Life Sciences (BOKU), Vienna, Austria, is an expert in plant functional genomics, genetic populations and molecular marker development. As Principal Investigator, she coordinated the efforts to analyze several hundred transgenic stone fruits and grapevine plants. She was responsible for the genetic characterization of the largest collection of apricot accessions, leading to a molecular conformation of the geographic spread of apricot. She has contributed to allergen research in fruits, with the aim of developing improved detection methods for traces of food allergens in fresh and processed plant-derived products. She is also responsible for a new research area involving important bioenergy plants, for example, *Jatropha curcas*. Since this non-edible plant is at a non-domesticated level, it requires the development of new 'OMICs' tools for its breeding and selection.

Dr Gaetano Perrotta has about 15 years' experience in genomics, biotechnology and molecular biology of higher plants. Major research interests focus on the study of genes involved in photo-perception of higher plants, fruit ripening and the metabolic pathways which control, either directly or indirectly, the biosynthesis and the accumulation of key metabolites. Some of his related research topics that have been developed in recent years concern the study of genes involved in light perception and flavonoid biosynthesis. He has managed a number of research projects funded by national and international institutions and has published more than 30 papers in peer-reviewed journals.

Dr Palmiro Poltronieri is a researcher at the Agrofood Department of the Italian National Research Council. He is co-founder of Biotecgen SME – a service company involved in European projects such as the FP VI project, 'Novel roles for non-coding RNAs', RIBOREG. He holds a PhD in Molecular and Cellular Biology from Verona University, and from 1996 to 1997 was post-doctoral fellow funded by the Japanese Society for the Promotion of Science at Tsukuba University. Since 1999, as a researcher for the NRC, he has been studying plant protease inhibitors and their applications. Presently, he applies molecular methods based on DNA arrays and protein chip tools to focused research topics. Current interests are water stress response in roots of tolerant and sensitive chickpea varieties, and the activation of the jasmonic acid synthesis pathway. He is an associate editor of *BMC Research Notes*.

Dr Md Abdur Rahim was born in 1980 in Rangpur, Bangladesh. He completed a bachelor's degree (BScAg(Hons)) in Agriculture and a Master's degree in Genetics and Plant Breeding at the Sher-e-Bangla Agricultural University, Dhaka, Bangladesh in 2004 and 2006, respectively. His Master's thesis was on the genetic diversity of rice. He is working as an Assistant Professor in the Department of Genetics and Plant Breeding at the same university, and is responsible for teaching and research on Plant Breeding & Genetics. He has been continuing his PhD with a CARIPARO fellowship at the Doctoral School of Crop Science, University of Padova, Italy, since 2011. Currently, he is working on transcriptomics in peach and pear fruit ripening.

Dr Angelo Santino holds a bachelor's degree in Biology. He was a fellow at the Institute of Natural Resources, Chatham Maritime, 1992, and visiting scientist in 1999 at the John Innes Centre, Norwich, UK, project title: 'Molecular cloning and characterisation of plant lipoxygenases from nutty species'. In 2004, he was at the Institut de Biologie Moléculaire des Plantes-CNRS, Strasburg, France, on the project: 'Phytoxylipins to improve the safety of plant products' (CNR-CNRS project in collaboration with Prof. E. Blee); a visiting scientist in 2005 at the John Innes Centre, Norwich, UK, working with Prof. R. Casey, on the 'Production of new biocatalysts for natural flavour and aroma synthesis' (CNR short-term fellowship); a visiting scientist in 2007 at the Institut des Sciences du Végétal, CNRS, Gif sur Yvette, France, on the project: 'Investigating the role of oxylipins in legume root development and nodulation', a collaboration that sent his PhD student for three months to work in Martin Crespi's laboratory. He is a member of the Euro Fed Lipid Association and the Italian Society of Agriculture Genetics (SIGA). He has published in top journals on plant lipoxygenases and on the enzymes in the oxylipin synthesis pathway.

Dr Marco Taurino graduated in 2004 in Biological Science at the University of Lecce, defending a thesis on plant cell wall biochemistry. He has a master's in 'Bioindustrial research in natural antioxidant extraction' at the National Research Council (CNR-ISPA) in Lecce. He obtained his PhD in plant physiology (on plant–pathogen interactions and the oxylipins pathway) at the National Center of Biotechnology of CSIC, Madrid (Universidad Autónoma de Madrid) with special competence in biochemistry and molecular biology. Now he has a regional fellowship to develop his project at CNR-ISPA Lecce, as a member of the group of Dr Giovinazzo, on the production of resveratrol and its derivates from grapevine cell culture.

Dr Raffaele Testolin is Professor of Fruit Science at the University of Udine, Italy, and lecturer on 'Fruit Science' and 'Genetic Resources in Agriculture'. Trained as a horticulturist, in the 1990s he shifted his interest to fruit molecular genetics, developing pioneering studies on the development of DNA microsatellite markers in fruit crops and their use in the genetic analysis of germplasm collections, linkage maps and mapping genes of agronomic interest, such as dioecy, anthocyanin biosynthesis, and resistance to plant diseases. He currently manages breeding programs in kiwifruit, grape and apple. His research has been funded by national and European authorities as well as by private companies. He is co-founder and President of the Institute of Applied Genomics, a not-for-profit research centre devoted to the structural and functional analysis of genomes (www.appliedgenomics.org). He has authored over 120 research papers, book chapters and plant patents.

Dr Livio Trainotti, born in Rovereto, Italy, in 1968, has been an Associate Professor of Botany in the Department of Biology of the University of Padova since 2006. After an MS degree in Biology (1992), he spent one year at the 'Institut für Pflanzen Genetik und Kultur Pflanzen Forschung' (IPK, Gatersleben, Germany). At the University of Padova he received a PhD in Evolutionary Biology (1997) and became a lecturer in Plant Physiology (1998). His main research interests deal with the genetic control of fruit ripening, investigated by means of genomics tools, with microarray platforms developed in his laboratory, and the manipulation of the genomes of tomato, tobacco and Arabidopsis in order to characterize the functions of genes of model and non-model species (such as peach and strawberry).

Introduction

This book aims to provide an overview of research advances in plant genomics, functional genomics and plant phenotyping, exploring the next generation technologies (Chapter 1), small RNAs and RNA silencing (Chapters 2 and 7), epigenetics (Chapter 3), metabolomics (Chapters 4 and 5), transcriptomics and functional genomics in conifers (Chapter 5), in tomato (Chapter 6), with a special focus on the interactions between hormones and light response genes, in grape (Chapters 7 and 8) and in peach (Chapter 9), doubled haploid technology in breeding of *Brassica napus* (Chapter 10), biotechnological approaches in cereal crops (Chapter 11) and biotechnological approaches to the production of bioactives such as resveratrol in biofermentors and in modified tomato plants (Chapter 12).

Plant functional genomics is presented under different approaches, in crop plants (Chapter 1) and in conifers (Chapter 5), with summaries of studies in genomics and transcriptomics in berries and fruit trees (Chapters 7 and 8). These completed scientific advancements have the potential to improve specialty crops (fruits, vegetables) and other plants for food and non-food applications.

In the coming years, these technologies will influence the scientific advances with applicable uses, especially in such fields as agronomy, stress-resistant varieties, improvement of plant fitness, improving crop yields, and other non-food applications. Skilled human resources are an essential building block for competitiveness. Supporting the training and the acquisition of expertise, of young scientists in particular, will help to widen their skill base and to develop links within and between the academic and industrial research environments.

This book is aimed not only at plant scientists but also at academic staff and students, thanks to the involvement of several authors with international genome sequencing projects and functional genomics (the *Solanaceae* (tomato and potato), *Rosaceae* (strawberry, peach), grape, and conifers genomics groups). In the topics covered, differences in genome structure, organization, small RNAs, and types of fruit are presented and discussed from the point of view of different species and groups, benefiting both plant students and specialists focused on individual plants.

In several chapters, ongoing and former international projects are presented, together with new approaches and technologies, often led by private companies, to produce novel tree varieties and tree transgenics. Furthermore, the book aims to focus the attention of the public authorities and the scientific community on the problematics and the monitoring of trials with new transgenic plants, providing some links and websites to monitor the activities of specific European Cooperation in Science and Technology (COST) actions and international research projects.

From plant genomics to -omics technologies

Palmiro Poltronieri, Institute of Sciences of Food Productions, ISPA-CNR, Italy

DOI: 10.1533/9781908818478.3

Abstract: The new technologies for DNA and RNA sequencing have made it possible to identify and analyse individual transcripts and differentially spliced isoforms at a relatively low cost. Several plant genomes are now known, as is the genome output: a multitude of transcripts. Studies are ongoing to understand what their function is and to locate their products in networks and pathways. One useful tool applied for this purpose is RNA silencing. Antisense RNA is one of the gene knock-out technologies that are used in cell and tissue cultures as well as in whole plants. Novel applications and the production of transgenic plants and trees require surveillance of the positive and negative effects on the environment and on life biodiversity in general that may result from their introduction.

Key words: RNA, expressed sequence tags (ESTs), serial analysis of gene expression (SAGE), cDNAs, cap analysis of gene expression (CAGE), transcriptomics, next generation sequencing (NGS), proteomics, metabolomics, systems biology, targeting induced local lesions in genomes (TILLING), quantitative trait locus (QTL), functional genomics, RNA silencing.

At the beginning of the last decade a revolution in high-throughput DNA sequencing, based on a high number of capillaries, high automation, robotic handling and microplate preparation of copy DNAs (cDNAs), allowed the collection of complete transcribed sequences (full-length cDNA libraries) (Carninci et al. 2003) to be produced for the human and mouse complete genomes. This approach was immediately applied to model plants such as

Arabidopsis. The general opinion was that DNA coding sequences were present in genomes in relatively small numbers (approx. 25 000–30 000 genes), surrounded by junk non-coding sequences. Since 2003, large-scale studies have aimed at the identification of RNA transcripts non-coding for proteins (ncRNAs), as a massive output of transcription (Frith et al. 2006). RNomics took the stage from the hypothesis that DNA is the static element of genetic information, the hardware, while RNA is the active information, the software. The ability of RNAs to orchestrate chromatin states, DNA transcription, differential splicing, RNA translation, post-transcriptional modification and protein stability determines a hidden layer of complexity of genetic information. In this way, not only did protein product numbers increase through the use of different transcription starts and different splicing sites, but also genes producing regulatory RNAs (long and small RNAs) were taken into account.

ENOD40 (Campalans et al. 2004), one of the first plant riboregulators, is produced in legume roots during nodulation, inducing the reprogramming of legume root cells. It functions as a structured RNA, through its binding to an RNA binding protein, but is also transcribed into a small peptide, thus belonging to a novel category of plant dual RNAs (Bardou et al. 2011).

In the years 2004 to 2007, the 'Riboreg' project, under the auspices of the European Commission VIth Framework Programme (FP6), brought together scientists such as Martin Crespi, Hervé Vaucheret, Joszef Burgyan, Javier Paz-Ares, Sakari Kauppinen and Jean-Marc Deragon, focused on RNAs in plants (Poltronieri and Santino 2012). A cooperative activity supported the production of the first Arabidopsis DNA microarray targeting ncRNAs and RNA binding proteins, to study transcripts in different tissues and in legume plants subjected to environmental stresses (Laporte et al. 2007).

DNA microarray platforms are today available for many plants, from Affymetrix GeneChip technology to in-situ synthesised microarrays (CombiMatrix, NimbleGen, Oxford Gene Technology, Agilent). Nowadays, transcriptomic studies are possible even for plants without sequenced genomes, through the development of expressed sequence tag (EST) libraries, cDNA collections, and high-throughput transcript profiling and next-generation sequencing (NGS). The new sequencing technologies (454/Roche GS FLX, SOLiD, Illumina GAIIX, new NGS platforms) have set up the basis for genome-wide comprehensive transcriptomics and analysis of RNAs (Metzker 2009; Oshlack et al. 2010). One recent effort focused on the transcriptome of *Catharanthus roseus* (SmartCell EU project, *http://www-smart-cell.org*).

1.1 SuperSAGE

Serial analysis of gene expression (SAGE) was a method exploited during the pre-genomic era to individuate each transcript based on sequencing short tags of 16 to 18 bases in length. Nowadays SuperSAGE, based on longer reads, such as tags of 26 bases in length, allows the powerful serial analysis of gene expression (Matsumura et al. 2012) (Figure 1.1).

Several protocols exploit restriction enzymes releasing 26-mer nucleotides and the application of tags to recognise the 5' and 3' ends of sequenced nucleic acids, allowing whole transcriptome studies to be performed. High-throughput SuperSAGE or DeepSuperSAGE is based on massive sequence analysis on the new, high-throughput NGS platforms. HT-SuperSAGE is suitable for use with the Illumina Genome Analyzer and the SOLiD sequencer (Matsumura et al. 2010). This approach allows deep transcriptome analysis and multiplexing, while reducing the time, cost and effort of the analysis.

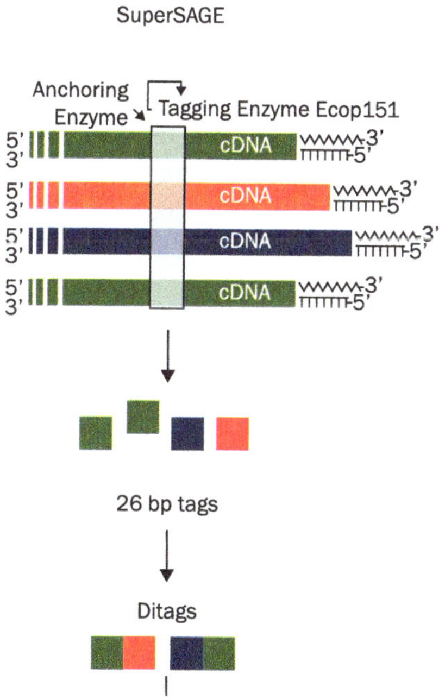

Figure 1.1 SuperSAGE protocol scheme: cDNA cleavage, formation of DNA tags and ligation into ditags

Source: GenXpro, Germany

SuperSAGE-Arrays were used for high-throughput transcription profiling studies, in legume genomes without completed DNA sequencing, to find differentially expressed genes in stress-tolerant and stress-susceptible genotypes. The DeepSuperSAGE protocol applied to chickpea (Molina et al. 2008, 2011) allowed early global transcriptome changes in drought- and salt-stressed chickpea roots and nodules to be detected and metabolic pathways relevant to these stress responses to be identified. Different varieties responded differently to abiotic stresses, showing significant intra-specific genetic variability. These studies identified new up-regulated and down-regulated genes and isoforms with tissue-specific expression, and assigned gene function through homology with genes already present in databases.

The SuperSAGE data were used to design Taqman primers splice variant-specific and isoform-specific for chickpea genes in real-time polymerase chain reaction (PCR) studies in root tissues of two not yet studied varieties in drought stress conditions, validating these data with high performance liquid chromatography (HLPC) and mass spectrometry analysis of the intermediates produced and active hormone forms (De Domenico et al. 2012).

Recent sequencing technologies have revitalised sequencing approaches in genomics and have produced opportunities for various emerging analytical applications. A new European FP7 project, AB-Stress, based on DeepSuperSAGE massive analysis of cDNA ends (MACE), (Kahl et al. 2012), epigenetics and DNA methylation changes, aims to elucidate the stress-induced small RNome in legumes (pea and Medicago) plants during biotic and abiotic stresses (Poltronieri and Santino 2012).

1.2 CAGE – cap analysis of gene expression

DNA Next-Generation Sequencing technologies have also been applied to the identification of the 5' end of cDNAs and to the differentiation of transcription start sites of expressed genes.

CAGE is a 5' sequence tag technology applied to globally determine transcription start sites in the transcribed genome and to measure the expression levels: the production of tags is combined with Next-Gen sequencing for high-throughput processivity (Takahashi et al. 2012). In principle, a CAGE protocol resembles a SuperSAGE protocol, except for the selective capturing of 5' capped mRNAs, a method previously exploited by Carninci in the preparation of full-length cDNA libraries.

Recently, CAGE has been adapted to the HeliScope single molecule sequencer. Despite significant simplifications in the CAGE protocol, it is still a labour-intensive protocol (Itoh et al. 2012).

1.3 -Omics and new advances in plant functional genomics

Systems biology enables the determination of how the interconnected networks of genes and gene products work together in steering biological processes, for instance, to produce fruit and grain, or to determine the performance of the plant under different specific environmental conditions (Mochida and Shinozaki 2011). Systems biology will allow scientists to reveal how natural genetic variation creates biodiversity and, together with innovative genomic technologies, will support researchers in the discovery of methods for breeding plants.

There is a need for resources and analytical tools for functional genomics, through different approaches, for the application of 'omics' (transcriptomics, proteomics, metabolomics) technologies, plant phenotyping, Quantitative Trait Loci (QTL) analysis and identification of genes by expression QTL (eQTL) (Figure 1.2), in order to understand the molecular systems that regulate various plant functions.

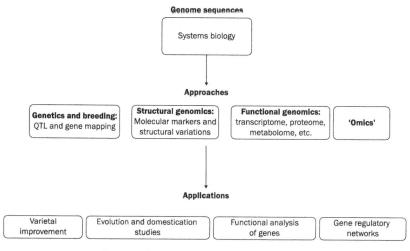

Figure 1.2 Approaches to applications of systems biology and -omics technologies in plant crops improvement

Great numbers of plant genomes have been released in recent years, exploited to enable an understanding of vascular plants (Banks et al. 2011) or highly important tree species (Grattapaglia et al. 2012; Slavov et al. 2012), phylogenetic studies or advances in phenotype analysis (Chia et al. 2012; The Tomato Genome Consortium 2012). A recent overview of plant genomes and their exploitation discusses this wealth of data on plant genomes (Ranjan et al. 2012). Traditional studies developing plant resources, such as conventional breeding and marker-assisted selection, need to be supported by the genomics and -omics information, thanks to the new high-throughput platforms. These efforts can produce improvements in food crops and non-food plants, obtaining an increase in the production of plants with desired traits.

TILLING (targeting induced local lesions in genomes) and collections of plant mutants, reverse and forward genetics (tissue-specific expression and gene silencing) have been extended from model plants (Arabidopsis, Medicago) to important food crops.

In tomato, the development of genetic and genomic resources has led to the development of functional genomic resources of tomato as a model cultivar with great importance for human nutrition (Ranjan et al. 2012; Matsukura et al. 2008). Tomato populations treated with 1.0% ethyl methanesulfonate (EMS) showed a frequency (one mutation per 737 kb) suitable to produce an allelic series of mutations in the target genes (Minoia et al. 2010; Okabe et al. 2011). Micro-Tom TILLING platforms were used for efficient mutant isolation, as a tool to study fruit biology and to obtain novel genetic material to be used to improve agronomic traits. A tomato in silico database, TOMATOMA, is a relational system interfacing modules between mutant line names and phenotypic categories (Saito et al. 2011).

Small RNAs include microRNAs, small interfering RNAs (siRNAs) and *trans*-acting siRNAs (tasi-RNAs) (Eamens et al. 2011; Poltronieri and Santino 2012). MicroRNAs target mRNAs by forming duplexes on the complementary seed sequences (seven bases in length) in mRNA transcripts. miRNAs negatively affect their targets through a variety of transcriptional and post-transcriptional mechanisms, such as mRNA degradation or blocking transcription.

RNA silencing and RNA interference allow specific knockdown of individual gene targets. At low concentration, microRNAs are able to affect the expression of several genes and of hundreds of mRNAs of one gene target. Because miRNAs target several different mRNA species, often in a tissue-specific manner, the delivery of RNAs complementary to miRNAs, as miRNA blockers, may affect and control cell growth more strongly than antisense RNA and RNA analogs. Hence, the ability of individual miRNAs to target multiple genes and pathways is potentially a major advantage. Several methods have been developed to inhibit a specific microRNA, such as target mimicry (Rubio-Somoza and Manavella 2011) or the siRNA sponges, in which long RNA strands containing hundreds of thousands to millions of nucleotides are designed to be cleaved by cells' RNA processing machinery into siRNAs inside the cells, to produce a high copy number of expressed antagomiRs (sequences that complex and sequester microRNAs) (Lee et al. 2012).

Diverse and complementary technologies to study plant adaptation in response to biotic and abiotic stress will benefit from top-down and bottom-up genomics approaches to identify potential gene candidates for innovative molecular breeding strategies. Gene overexpression and gene knock-out in plant tissue cultures (Ariel et al. 2010) and RNA interference will take the stage in coming years (Rubio-Somoza and Manavella 2011).

The exploitation of RNA silencing and antisense technologies to control gene expression has been translated into new plant phenotypes and tree populations with novel traits. Several international collaborations are at an advanced stage, for example, European COST activities FA0804: *http://molfarm.ueb.cas.cz/* 'Molecular farming: plants as a production platform for high value proteins', FA1006 http://www.plantengine.eu/ 'Plant metabolic engineering for high value products', and the EU collaborative project 'Green factories for the next generation of pharmaceuticals', SmartCell. Other projects aim to develop novel tree genetic strategies, such as the NovelTree project *http://cordis.europa.eu/projects/rcn/88733_en.html*, and to improve major forest genetics and forestry research infrastructures (Trees4Future, *http://www.trees4future.eu/*). There are two coordinated approaches to the topic of plant biotechnology: the first, COST FP0905, for monitoring transgenic trees in vitro and in field trials, 'Biosafety of forest transgenic trees: improving the scientific basis for safe tree development and implementation of EU policy directives', *http://www.cost-action-fp0905.eu/* (Walter et al. 2010); and the second COST FA0806, 'Plant virus control employing RNA-based vaccines', *http://costfa0806.aua.gr/*.

The exploitation of RNA silencing and antisense technologies to control gene expression has already been translated into new plant phenotypes and tree varieties adapted to cold climates (such as the SENESCO proprietary technology to produce transgenic plants and trees).

Recently Carol Auer summarised the state of the art of plant biotechnologies with a special focus on new approaches based on small

RNAs, RNA interference and the production of RNA-mediated traits in plants (Auer 2011). The potential of RNA-regulated traits in non-food plants and biofuel-producing plants is enormous. Accordingly, new methods for risk analysis are required to perform analyses of off-target effects and the persistence of RNAs in the environment.

Complexity Science, Informing Science, -Omics Engineering could effectively support and integrate the methodologies of different academic fields. To this end, cloud computing and the sharing of data networks are possible today using new tools and software, designed to work in Linux-based environments, while new systems will become available for use on Microsoft, such as the Windows2Galaxy project. The continued adoption of Galaxy by the life sciences community depends on the enhancement of features and the development of new functionality. A number of new features were recently highlighted by members of the Galaxy development team (Li et al. 2012).

Some recent scientific developments have shown an impact on food and non-food crops: breakthroughs in understanding how plant cells recognise different hormones and which signalling pathways are activated by hormones (Razem et al. 2006; Fujii et al. 2009; Yin et al. 2009) and the links between epigenetics and abiotic stress memory (Urano et al. 2010); the role of plant sRNAs and epigenetics in the regulation of development and stress response (Chuck and O'Connor 2010; Matsui et al. 2008; Merchan et al. 2007); and understanding how interactions between genome elements (DNA, RNA) and the environment make a plant body. Understanding how non-coding RNAs work will reveal novel mechanisms involved in growth control and differentiation.

Skilled human resources are an essential building block for competitiveness. Supporting young scientists and their training in new technologies will help to widen their skill base and to develop links within and between the academic and industrial research environments. In the forthcoming years these advances will support the production of plant varieties better suited to resist biotic and abiotic stresses, for food and non-food applications.

1.4 Bibliography

Ariel F, Diet A, Verdenaud M, Gruber V, Frugier F, et al. Environmental regulation of lateral root emergence in *Medicago truncatula* requires the HD-Zip I transcription factor HB1. *Plant Cell* 2010, 22:2171–83.

Auer C. Small RNAs for crop improvement. In Erdmann VA, Barciszewski J (Eds) *Non coding RNAs in plants*. Springer-Verlag, Berlin, 2011, pp. 461–84.

Banks JA, Nishiyama T, Hasebe M, Bowman JL, Gribskov M, et al. The *Selaginella* genome identifies genetic changes associated with the evolution of vascular plants. *Science* 2011, 332:960–3.

Bardou F, Merchan F, Ariel F, Crespi M. Dual RNAs in plants. *Biochimie* 2011, 93(11):1950–4.

Campalans A, Condorosi A, Crespi M. *Enod40*, a short open reading frame-containing mRNA, induces cytoplasmic localization of a nuclear RNA binding protein in *Medicago truncatula*. *Plant Cell* 2004, 16:1047–59.

Carninci P, Waki K, Shiraki T, Konno H, Shibata K, et al. Targeting a complex transcriptome: the construction of the mouse full-length cDNA encyclopedia. *Genome Res.* 2003, 13:1273–89.

Chia J-M, Song C, Bradbury PJ, Costich D, de Leon N, et al. Maize HapMap2 identifies extant variation from a genome in flux. *Nat. Genet.* 2012, 44:803–7.

Chuck D, O'Connor G. Small RNAs going the distance during plant development. *Curr. Op. Plant Biol.* 2010, 13:40–5.

De Domenico S, Bonsegna S, Horres R, Pastor V, Taurino M, et al. Transcriptomic analysis of oxylipin biosynthesis genes and chemical profiling reveal an early induction of jasmonates in chickpea roots under drought stress. *Plant Physiol. Biochem.* 2012, 61:115–22.

Eamens AL, Curtin SJ, Waterhouse PM. The *Arabidopsis thaliana* double-stranded RNA binding (DRB) domain protein family. In: Erdmann VA, Barciszewski J (Eds) *Non coding RNAs in plants*. Springer-Verlag, Berlin, 2011, pp. 385–405.

Frith MC, Bailey TL, Kasukawa T, Mignone F, Kummerfeld SK, et al. Discrimination of non-protein coding transcripts from protein-coding mRNA. *RNA Biol.* 2006, 3:40–8.

Fujii H, Chinnusamy V, Rodrigues A, Rubio S, Antoni R, et al. In vitro reconstitution of an abscisic acid signalling pathway. *Nature* 2009, 462:660–4.

Grattapaglia D, Vaillancourt R, Shepherd M, Thumma B, Foley W, et al. Progress in Myrtaceae genetics and genomics: Eucalyptus as the pivotal genus. *Tree Genetics & Genomes* 2012, 8(3):463–508.

Itoh M, Kojima M, Nagao-Sato S, Saijo E, Lassmann T, et al. Automated workflow for preparation of cDNA for cap analysis of gene expression on a single molecule sequencer. *PLoS One* 2012; 7:e30809.

Kahl G, Molina C, Rotter B, Jungling R, Frank A, Krerdom N, Hoffmeier K, Winter P. Reduced representation sequencing of plant stress transcriptomes. *Journal of Plant Biochemistry Biotechnol.* 2012, 21(15):119–27.

Laporte P, Merchan F, Amor BB, Wirth S, Crespi M. Riboregulators in plant development. *Biochem. Soc. Trans.* 2007, 35:1638–42.

Lee JB, Hong J, Bonner DK, Poon Z, Hammond PT. Self-assembled RNA interference microsponges for efficient siRNA delivery. *Nature Materials* 2012, 11:316–22.

Li P, Goecks J, Lee T-L. Turning pipe dreams into reality. *Genome Biology* 2012, 13:318.

Matsui A, Ishida J, Morosawa T, Mochizuki Y, Kaminuma E, et al. Arabidopsis transcriptome analysis under drought, cold, high-salinity and ABA treatment conditions using a tiling array. *Plant Cell Physiol.* 2008, 9:1135–49.

Matsukura C, Aoki K, Fukuda N, Mizoguchi T, Asamizu E, et al. Comprehensive resources for tomato functional genomics based on the miniature model tomato micro-tom. *Curr. Genomics* 2008, 9:436–43.

Matsumura H, Yoshida K, Luo S, Kimura E, Fujibe T, et al. High-throughput SuperSAGE for digital gene expression analysis of multiple samples using next generation sequencing. *PLoS One* 2010, 5:e12010.

Matsumura H, Urasaki N, Yoshida K, Krüger DH, Kahl G, Terauchi R. SuperSAGE: powerful serial analysis of gene expression. *Methods Mol. Biol.* 2012; 883:1–17.

Merchan F, de Lorenzo L, Rizzo SG, Niebel A, Manyani H, et al. Identification of regulatory pathways involved in the reacquisition of root growth after salt stress in Medicago truncatula. *Plant J.* 2007, 51:1–17.

Metzker ML. Sequencing in real time. *Nat. Biotechnol.* 2009, 27:150–1.

Minoia S, Petrozza A, D'Onofrio O, Piron F, Mosca G, et al. A new mutant genetic resource for tomato crop improvement by TILLING technology. *BMC Res. Notes* 2010, 3:69.

Mochida K, Shinozaki K. Advances in omics and bioinformatics tools for systems analyses of plant functions. *Plant Cell Physiol.* 2011, 52:2017–38.

Molina C, Rotter B, Horres R, Udupa SM, Besser B, et al. SuperSAGE: the drought stress-responsive transcriptome of chickpea roots. *BMC Genomics* 2008, 9:553.

Molina C, Zaman-Allah M, Khan F, Fatnassi N, Horres R, et al. The salt-responsive transcriptome of chickpea roots and nodules via DeepSuperSAGE. *BMC Plant Biol.* 2011, 11:31.

Okabe Y, Asamizu E, Saito T, Matsukura C, Ariizumi T, et al. Tomato TILLING technology: development of a reverse genetics tool for the efficient isolation of mutants from Micro-Tom mutant libraries. *Plant Cell Physiol.* 2011, 52:1994–2005.

Oshlack A, Robinson MD, Young MD. From RNA-seq reads to differential expression results. *Genome Biology* 2010, 11:220.

Poltronieri P, Santino A. Non-coding RNAs in intercellular and systemic signaling. *Front. Plant Sci.* 2012; 3:141.

Ranjan A, Ichihashi Y, Sinha NR. The tomato genome: implications for plant breeding, genomics and evolution. *Genome Biology* 2012, 13:167.

Razem FA, Baron K, Hill RD. Turning on gibberellin and abscisic acid signaling. *Curr. Op. Plant Biol.* 2006, 9:454–9.

Rubio-Somoza I, Manavella PA. Mimicry technology: suppressing small RNA activity in plants. *Methods Mol. Biol.* 2011, 732:131–7.

Saito T, Ariizumi T, Okabe Y, Asamizu E, Hiwasa-Tanase K, et al. TOMATOMA: a novel tomato mutant database distributing Micro-Tom mutant collections. *Plant Cell Physiol.* 2011, 52(2):283–96.

Slavov GT, Difazio SP, Martin J, Schackwitz W, Muchero W, et al. Genome resequencing reveals multiscale geographic structure and extensive linkage disequilibrium in the forest tree *Populus trichocarpa*. *New Phytol.* 2012, 3 August. doi: 10.1111/j.1469-8137.2012.04258.x. [Epub ahead of print]

Takahashi H, Lassmann T, Murata M, Carninci P. 5' end-centered expression profiling using cap-analysis gene expression and next-generation sequencing. *Nat. Protoc.* 2012, 7:542–61.

Tardieu F, Tuberosa R. Dissection and modelling of abiotic stress tolerance in plants. *Curr. Op. Plant Biol.* 2010, 13:206–12.

The Tomato Genome Consortium. The tomato genome sequence provides insights into fleshy fruit evolution. *Nature* 2012, 485:635–41.

Urano K, Kurihara Y, Seki M, Shinozaki K. 'Omics' analyses of regulatory networks in plant abiotic stress responses. *Curr. Op. Plant Biol.* 2010, 13:132–8.

Walter C, Fladung M, Boerjan W. The 20-year environmental safety record of GM trees. *Nat. Biotechnol.* 2010, 28:656–8.

Yin P, Fan H, Hao Q, Yuan X, Wu D, et al. Structural insights into the mechanism of abscisic acid signaling by PYL proteins. *Nat. Struct. Mol. Biol.* 2009, 16:1230–6.

Plant microRNAs

Moreno Colaiacovo and Primetta Faccioli, Consiglio per la Ricerca e la sperimentazione in Agricoltura, Centro di ricerca per la genomica e la postgenomica animale e vegetale, Fiorenzuola d'Arda, Italy

DOI: 10.1533/9781908818478.15

Abstract: Plants show a remarkable developmental plasticity to adapt their growth to changing environmental conditions. Recently small non-coding RNAs and microRNAs have been identified by their role in the regulation of gene expression and functions in development and in response to the environment. Plant microRNAs have been grouped into ancient and recently formed miRNAs. The ancient types are highly conserved in the plant kingdom. The oldest miRNAs tend to be highly expressed in plant cells and usually perform more fundamental tasks, whereas young miRNAs are often less abundant and are induced by specific conditions.

Plant miRNAs target only a small number of mRNAs; however, by largely targeting transcription factor genes, they act in key biological processes, performing metabolic functions, responding to abiotic and biotic stresses, and guiding developmental trajectories. Moreover, miRNAs are regulators of plant–microbe interactions, and several miRNAs move through the phloem from leaves to roots as effectors of plant response to environment factors. The ability to regulate plant microRNA functions opens new perspectives in the field of crop improvement.

Key words: microRNAs, mRNA targets, evolution, conserved miRNAs, species-specific miRNAs, development, stress, competing endogenous RNAs, artificial microRNAs.

2.1 Introduction

MicroRNAs (miRNAs) are short non-translated RNAs that are widespread in all branches of eukaryotic life, from single-cell organisms, such as the green algae *Chlamydomonas reinhardtii*, to multicellular organisms, such as animals and plants.[1-8] Initially their discovery in different organisms was associated in every case with their role in developmental patterning, since their first identification in the nematode *Caenorhabditis elegans*.[9-11] This fact led to the idea that miRNAs might have evolved as a consequence of multicellularization, but in 2007 their identification in *C. reinhardtii*[7,8] refuted this theory. Additionally, it was found that miRNAs perform a wide range of biological activities in eukaryotic cells, besides developmental regulation.[12-23]

2.2 Transcription of miRNA genes

Plant miRNAs arise from genetic loci known as MIRNA genes, which exist mostly as independent transcription units. miRNAs are transcribed as longer molecules (primary miRNAs, or pri-miRNAs) by RNA polymerase II. The role of this enzyme in miRNA biogenesis is supported by several lines of evidence: first, plant pri-miRNAs have a 5′ cap and a 3′ poly (A) tail, which are typically found in pol II transcripts; second, the promoter regions of MiRNA genes usually contain features which are characteristic of pol II genes, such as TATA box, transcription factor binding sites and transcription initiator (INR) elements.[24,25] In most cases, plant miRNAs are located in intergenic regions, but some exceptions to this rule have recently been found.[26] Some examples of intronic miRNAs (mirtrons) have been reported in Arabidopsis (miR402 and miR838)[27,28] and rice (miR1429.2),[29] while an exonic miRNA was found in rice (miR3981).[30] Although the majority of plant pri-miRNAs give rise to a single mature miRNA, polycistronic loci have also been reported, in which the pri-miRNA can form more than one hairpin, each containing a distinct mature miRNA.[31]

2.3 MicroRNA processing

The mechanism by which pri-miRNAs take the route of the miRNA pathway instead of the translation machinery has not been completely elucidated; however, the general framework of miRNA biogenesis has been revealed.[6] After transcription, pri-miRNAs are involved in three processes: 5′ capping, splicing and polyadenylation at the 3′ tail (Figure 2.1). Several studies have shown that a key element in this

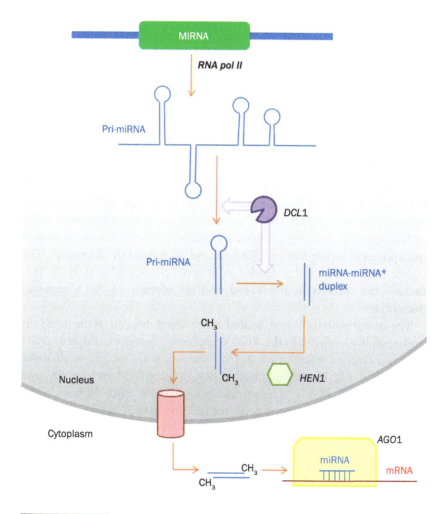

Figure 2.1   miRNA biogenesis in plant cells

phase is the nuclear cap-binding complex (CBC): when this complex is disrupted in loss-of-function mutants, cells accumulate unspliced pri-miRNA transcripts.[32-34] Another key element in this initial phase is SE (SERRATE) protein, a C2H2-type zinc-finger domain-containing protein.[35]

MiRNA processing from pri-miRNA to the mature miRNA requires DCL1, a dicer-like enzyme belonging to the RNAse III-like protein family. The dicer proteins contain different domains which are fundamental for the correct processing of pri-miRNAs: a helicase domain at the N-terminus, a PAZ domain in the middle and the catalytic RNase III domains with a double-stranded (ds)RNA-binding motif at the C-terminus. The processing releases a double-strand short RNA, whose length is determined by the distance between the PAZ domain and the RNase III domains.[36] In this process, DCL1 is helped by other proteins such as HYL1, a dsRNA-binding protein, and SE, which probably facilitate the positioning of DCL1 on the transcript.[37-39]

The DCL1 complex releases a small RNA duplex consisting of a miRNA and a miRNA* with a 2-nt overhang at the 3'-end. Then a small RNA methyltransferase (HEN1) adds a methyl group to the 2'-hydroxyl group of the 3'-terminal nucleotide in each strand, a modification which is thought to stabilize the molecule. Finally, the mature miRNA is loaded into an ARGONAUTE (AGO)-containing complex, known as RISC (RNA-induced silencing complex). Usually, only the mature strand is incorporated, while the miRNA* strand is ultimately degraded. The choice of which strand is loaded is believed to depend mainly on two factors: the stability at the 5'-end and the identity of the 5'-terminal nucleotide.

The strand with its 5'-end located at the less stable end of the duplex is preferentially loaded into the AGO complex;[2,40] moreover, different AGO enzymes tend to incorporate strands starting with specific nucleotides (AGO1 prefers a 5'-terminal uridine, AGO2 and AGO4 prefer a 5'A, and AGO5 prefers a 5'C).[41-43] The vast majority of plant miRNAs have a 5'U, which is in perfect agreement with the fact that AGO1 is the most active member of the AGO family in the miRNA pathway.

2.4 Modes of action

In plants, miRNAs regulate the expression of protein-coding genes mainly by cleavage of the corresponding mRNA targets, even if some lines of evidence suggest that translational repression is also possible.[44] Besides negative regulation of protein-coding mRNAs, some miRNAs have shown peculiar functions, such as triggering the biogenesis of trans-acting siRNAs (tasiRNAs), another small RNA species which in turn is able to negatively regulate other mRNAs.[45–47] Moreover, some miRNA variants tend to associate with AGO proteins that are involved in RNA-directed DNA methylation.[48] Therefore, miRNA-mediated regulation encompasses almost all levels of regulation in plant cells. More and more studies suggest a pervasive role of miRNAs in plant biology; their ability to move in surrounding tissues, thus participating in cell-to-cell signaling, was recently demonstrated.[49]

2.5 Evolution of miRNA genes

Plant miRNAs can be roughly divided into two main groups: ancient and young miRNAs. Ancient miRNAs are deeply conserved in the plant kingdom: all the angiosperm lineages share 21 miRNA families, and eight of them can be found in all common ancestors of the embryophytes. Young miRNAs are family or even species-specific: MIR472 is present in all core rosids, and at least nine families are known to have arisen in monocots.[5] The oldest miRNAs tend to be highly expressed in plant cells and usually perform more fundamental tasks, whereas young miRNAs are often less abundant and induced by specific conditions.[5] The vast majority of young miRNAs have few known functions, as they tend to lack targets, and they are processed less precisely by the miRNA machinery.[4] Although other theories exist, plant miRNAs are thought to have evolved by inverted duplication of protein-coding genes. After the initial duplication, an accumulation of mutations in the foldback arms can eventually lead to an increased affinity for the miRNA pathway; in this way a potential new miRNA may arise and be incorporated, in rare instances, into a new or existing regulatory network.[5] The theory of origin by inverted duplication is supported by the fact that newer miRNAs tend to be more similar in sequence to their target, because of a more recent duplication of the target gene and therefore a low number of mutations having occurred since the event.[4] It is estimated that plant miRNAs are born and lost at a high frequency, and relatively few of the evolving miRNAs actually become functioning regulators; the evolution of young miRNAs is in fact considered to be neutral.[5,50]

2.6 Differences from animal miRNAs

Current lines of evidence suggest that miRNAs arose independently in plants and animals.[5] Several differences exist between the two groups, ranging from the genomic organization to the mechanism of action. As reported above, in plants most miRNA genes are transcribed as independent units, whereas in animals intronic or even exonic miRNAs are much more common, as well as clusters of miRNA genes. Other differences pertain to the structural features of pre-miRNAs: plant miRNAs are generally longer than their counterparts in animals, and their size is more variable, ranging from 70 to several hundreds of bases. Their biogenesis also has some key differences: in plants, pri-miRNAs are processed to miRNA duplexes by just one enzyme (DCL1), while in animals the process is performed in two separate steps by Drosha and Dicer enzymes, which act respectively in the nucleus and in the cytoplasm. In plants, the processing of pri- and pre-miRNAs happens in the nucleus, and only the duplex is exported to the cytoplasm. Another peculiar feature of plant miRNAs is the 3' methylation by HEN1. Finally, plant miRNAs require extensive pairing to their targets, and they act mainly by slicing the mRNA to which they bind; animal miRNAs, on the contrary, exhibit only partial complementarity to their targets and prefer translational repression. Additionally, miRNA target sites are mainly located in the 3' untranslated region (UTR) of mRNA in animals, whereas in plants they are found both in untranslated regions and in the coding sequence. Because of their extensive complementarity, each of the plant miRNAs has few targets and each target mRNA has few target sites; animal miRNA targets, on the other hand, often have many sites, and each miRNA regulates a wide set of targets in a subtler way than plant miRNAs.[1]

2.7 miRNA functions

In contrast to animal miRNAs, which regulate approximately 60% of protein-coding genes, plant miRNAs target only a small number of mRNAs (about 1% of protein-coding genes).[51] Nevertheless, by largely targeting transcription factor-coding genes, they act in key biological processes, performing metabolic functions, responding to abiotic and biotic stresses, and guiding developmental trajectories. Moreover, miRNAs are also regulators of plant–microbe interactions during nitrogen (N) fixation by *Rhizobium* and tumor formation by *Agrobacterium*.[21] Thus, the overall impact of miRNA-based regulation in plants is huge. Since their first discovery in *C. elegans*, miRNAs have been linked to developmental processes. This role has been highlighted also in plants, where they appear to perform fundamental functions in basically all developmental phases and tissues.[17] It is now well known that the interaction of miR156 and miR172 drives the progression from the juvenile to the adult stage of vegetative growth, and then promotes the reproductive phase (Figure 2.2). The two miRNAs show complementary expression patterns. During earlier phases, miR156 is highly expressed, and its targets (SEPALLATA-SPL genes) are inhibited; when the plant grows, its level decreases, thus promoting the expression of SPL proteins. SPL9 and SPL10 activate miR172, which in turn dampens the expression of transcription factors (such as AP2-like proteins) that normally repress flowering; their inhibition, together with a higher level of other SPL proteins, makes the plant competent to flower, and the transition to the reproductive phase can occur. It is interesting to

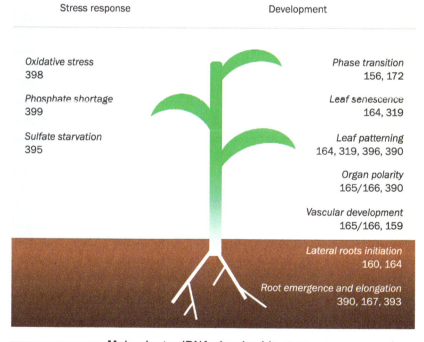

Figure 2.2 Main plant miRNAs involved in stress response and development

note that the floral patterning is also controlled by miR172 and its target AP2.[16,17] miRNAs are also deeply involved in other developmental processes. miR164 and miR319 act to delay leaf senescence, and they are also able to control the patterning of the leaves, together with miR396 and miR390. The miR165/miR166 family is involved in organ polarity with miR390 and in vascular development with miR159. Root architecture is also driven by a set of miRNAs: miR164 and miR160 play a role in the initiation of lateral roots, while miR390, miR167 and miR393 act during root emergence and elongation.[17] miRNA-mediated regulation of developmental processes is connected with other regulatory pathways of the plant through miRNA targets, which are often transcription factors. The way miRNAs affect development has been classified into three different classes: spatial restriction, buffering function and temporal regulation.[52] In the first case, miRNAs control developmental processes by reducing the expression level of their targets in the tissue where they accumulate, thus promoting cell differentiation. For example, miR165/166 follows this mode of action when regulating PHABULOSA (PHB) in Arabidopsis: this gene is polarly expressed on the adaxial side of leaves, while the miRNA accumulates on the adjacent abaxial domain. In the case of the buffering function, miRNAs keep the level of their targets on a well-defined expression domain: this action finely tunes the expression level of targeted genes. According to experimental evidence, it is suggested that miR164 controls the expression of *CUC2* (CUP-SHAPED COTYLEDON 2), and the same buffering function has been associated with miR156 and its target SPL9, as well as miR168 and its target AGO1. The third mode of action is temporal regulation, in which miRNAs act as developmental clocks. In Arabidopsis, the accumulation of miR172 during the vegetative phase induces a gradual decrease in the abundance of its target (AP2-like proteins); when the target reaches a critical threshold, it promotes flowering. Similarly, flowering time is also controlled by miR156 and its targets (SBP family transcription factors).

In agricultural systems, plant productivity is strongly influenced by environmental conditions. Drought, low temperature and salinity are the most important abiotic stress factors limiting crop productivity, together with biotic stress conditions due to pathogenic bacteria, fungi, viruses, insect pests and nematodes. Several findings have established the fundamental role of miRNAs in plant stress response, both to biotic and abiotic stress and to nutrient deprivation.[53] A direct link between miRNAs and stress response has emerged with the identification of targets for miRNA398, miR399 and miR395. Although plants have an efficient antioxidant system, some stress conditions, such as drought, cold, salinity, high light and heavy metals, can lead to an accumulation of Reactive Oxygen Species (ROS) in cells. Superoxide radicals (O_2^-) are produced by photo-reduction of oxygen in Photosystem I (PSI) of chloroplasts. In order to scavenge these radicals, plants rely on enzymes such as Cu-Zn superoxide dismutase 2 (CSD2), which is attached to the thylakoid of chloroplasts for a fast scavenging of superoxide radicals. Superoxide dismutase genes are up-regulated during oxidative stress; however, the up-regulation does not occur at the transcriptional level, but rather is caused by the dampening of a miRNA-mediated regulation. MiR398 targets both the plastidic CSD2 and the cytosolic CSD1, and keeps their transcripts at a low level during normal growth conditions; however, in response to oxidative stress, this miRNA is down-regulated and its targets are released for the scavenging of superoxide radicals. A further confirmation of this process came from transgenic plants: it was shown that plants carrying miR398-resistant mutations in their CSD2 mRNA were much more tolerant to diverse stress conditions than transgenic plants carrying the susceptible CSD2 gene.[54]

Referring to nutrient deprivation response as another example, phosphorus is an essential macronutrient for plants: it is a basic component of biological structures, and has a role in energy metabolism, signal transduction cascades and enzyme regulation. It is usually present at high levels in the soil, but the form of phosphorus available for uptake by plants – orthophosphate (P_i) – is far less abundant, and plants have therefore developed a set of adaptive responses to P_i shortage. In response to a reduced availability of P_i, many processes are activated which foster P_i acquisition through various regulatory routes.[55] The expression of P-starvation-responsive (PSR) genes has been shown to be regulated at a transcriptional level as well as at a post-transcriptional level, with several important transcription factors being identified, as well as the role of a microRNA (miR399), which was highlighted by Fujii et al. in 2005.[56] miR399 is one of the most interesting cases of miRNA-mediated stress response. Upon P_i starvation, miR399 is up-regulated until the re-addition of P_i to the soil. Moreover, Arabidopsis lines overexpressing miR399 show an increased P_i uptake, and in the case of P-sufficient conditions these plants show symptoms typical of P_i toxicity (chlorosis and necrosis at the tips of mature leaves), as well as an impairment in P_i allocation. The same effect was also seen in rice and tomato. In Arabidopsis there are three genes that are predicted to be targeted by miR399: a P_i transporter (PHT1;7), a DEAD box helicase, and a ubiquitin-conjugating E2 enzyme encoded by *UBC24*, which has been experimentally validated.

Consistently with this regulation, UBC24 shows high expression levels in the roots under P-sufficient conditions, whereas it is down-regulated in response to P starvation, with an expression pattern inverse to its regulator miR399. This miRNA-mediated regulation has been confirmed by many experimental studies, which have demonstrated that UBC24 transcript decreases in transgenic plants overexpressing miR399, but is unaffected in plants that overexpress a mutated miR399. Similarly, the deletion of the miRNA target site in the UBC24 transgene prevents the degradation of the transcript by miR399 during phosphate stress. Homologs of miR399 have been identified in rice, tomato, *Phaseolus vulgaris* and *Medicago truncatula*. In 2007, an interesting study was published demonstrating a novel mechanism for regulation of microRNA activity: target mimicry.[53] The authors investigated the relationship between miR399 and a family of non-coding RNAs showing a high level of complementarity to the mature sequence of this microRNA. This family (of which the main members in Arabidopsis are IPS1 and At4) had only short, non-conserved open reading frames (ORFs), but the motif complementary to miR399 was instead conserved in different plant species, thus suggesting a likely functional role. The experiments which were carried out allowed the authors to conclude that these non-coding RNAs can inhibit miRNA activity by sequestering the mature sequences to their complementary sites. However, because of mismatches in key positions of the site (10,11), the RNA is not cleaved by the silencing complex. This clever mimicry strategy allows the real target mRNAs to escape miRNA-mediated regulation, as was shown by overexpression experiments. The authors were also able to demonstrate that artificial target mimicry is a useful tool for performing functional analysis of plant miRNAs.

Interestingly, miR399 has been suggested to be a phloem-mobile long distance signal[57] responding to the need, during nutrient starvation, of a plant organ to communicate its requirements to other organs, communication that is probably mediated via phloem.[58] Long distance signaling is known to be fundamental in plants for the regulation of several processes, including leaf development, flowering and pathogen defense, and several microRNAs have been detected in the phloem sap of plant species.[59]

Sulfur is another indispensable nutrient for plants: only nitrogen, inorganic phosphate and potassium are more needed among the inorganic nutrients. Usually it is taken up by the roots as inorganic sulfate, and its lack leads to various physiological changes that allow more efficient sulfate acquisition and suspend sulfate assimilation. It has been shown that a miRNA, namely miR395, performs an important role in the response to sulfate starvation. First, it regulates ATP sulfurylases (APS1, APS3 and APS4), enzymes that catalyze the first step of the assimilation pathway; second, it targets a sulfate transporter (AST68) which is involved in the translocation of sulfate from roots to shoots. Expression analyses confirmed the induction of miR395 under conditions of low sulfate, whereas the APS1 transcript levels decrease. The opposite situation occurs when sulfur levels are sufficient.[14]

Interestingly, the expression profiles of several miRNAs involved in plant growth and development are significantly altered during stress. These findings imply a control by stress-responsive miRNA of the attenuation of plant growth and development under stress that is strictly related to auxin perception and signaling.[60]

2.8 The potential roles of microRNAs in crop improvement

miRNA-based technology, developed with the final aim of providing powerful functional genomics tools for the study of miRNA activity, can also be used for the regulation of the function of miRNAs themselves. Because of miRNAs' ability to silence specific genes, miRNA-based manipulation has emerged as a promising new tool for crop improvement via breeding strategies and genetic modification of agronomic traits (i.e. tolerance to environmental stress, high yield). Artificial miRNAs can be used as dominant suppressors of the activity of specific genes when brought inside the plant,[61] thus leading to the direct molecular modulation of plant traits. Amaras are based on the expression of endogenous miRNA precursor genes that have been manipulated to exchange the miRNA/miRNA* sequence, thus generating mature amines with high specificity for a target gene of interest and inducing its silencing without affecting the expression of other genes. Amaras have been used successfully as expression cassettes for gene silencing in Arabidopsis, rice,[61] tobacco and tomato,[62] and to target even genes that are not natural targets of natural miRNAs.[63] This approach can thus also be very useful for the validation of the function of genes putatively involved in interesting agronomical traits, assisting in their practical usage for crop improvement. Web-based miRNA designer software is available for the design of a miRNA-like structure for gene silencing in several species (*http://wmd3. weigelworld.org/cgi-bin/webapp.cgi*).

In vitro chemically modified miRNA inhibitors, namely antagomiRs, can be used to silence endogenous microRNAs.[64] Based on antisense strategy, antagomiRs act as competitors for endogenous miRNA targets. The operating principle is thus the same as 'target mimicry'. Target mimicry, as reported above for P_i homeostasis, consists of a non-cleavable RNA that interacts with a complementary miRNA, thus inhibiting its activity. These 'competing endogenous RNAs' (ceRNAs) thus define a new layer of regulation of miRNA activity.[65] Artificial target mimicry may provide a powerful tool to manipulate the level of endogenous miRNA. A collection of transgenic plants expressing artificial target mimics has been generated in Arabidopsis to reduce the activity of most of the known miRNA families.[66] An engineered IPS1 has been proved to successfully inhibit the activities of miR156 and miR319 in Arabidopsis.[67] However, challenges are posed by the possibility that a single miRNA could interact with more than one gene, for example, different members of a gene family, thus leading to high expression levels of the whole gene family and not of a single gene. MicroRNA sponges represent another emerging powerful tool for miRNA loss of function studies that are based on the same principle of target mimicry. MicroRNA sponges are transcripts with repeated miRNA antisense sequences with a bulge or a mismatch at the cleavage site that can sequester miRNA from its endogenous target.[68] Sponges can adsorb a high level of complementary miRNAs, thus releasing the repression of *bona fide* mRNA targets in the cell.

2.9 Bibliography

1. Axtell, MJ, Westholm, JO and Lai, EC. Vive la différence: biogenesis and evolution of microRNAs in plants and animals. *Genome Biology* 12(4), 221 (2011).
2. Jones-Rhoades, MW, Bartel, DP and Bartel, B. MicroRNAs and their regulatory roles in plants. *Annual Review of Plant Biology* 57, 19–53 (2006).
3. Voinnet, O. Origin, biogenesis, and activity of plant microRNAs. *Cell* 136, 669–87 (2009).
4. Tang, G. Plant microRNAs: an insight into their gene structures and evolution. *Seminars Cell and Developmental Biology* 21, 782–89 (2010).
5. Cuperus, JT, Fahlgren, N and Carrington, JC. Evolution and functional diversification of MIRNA genes. *The Plant Cell* 23, 431–42 (2011).
6. Xie, Z, Khanna, K and Ruan, S. Expression of microRNAs and its regulation in plants. *Seminars Cell and Developmental Biology* 21, 790–7 (2010).
7. Molnár, A, Schwach, F, Studholme, DJ, Thuenemann, EC and Baulcombe, DC. miRNAs control gene expression in the single-cell alga *Chlamydomonas reinhardtii*. *Nature* 447, 1126–9 (2007).

8. Zhao, T, Li, G, Mi, S, Li, S, Hannon, GJ, et al. A complex system of small RNAs in the unicellular green alga *Chlamydomonas reinhardtii*. *Genes and Development* 21, 1190–203 (2007).
9. Lee, RC, Feinbaum, RL and Ambros, V. The C. elegans Heterochronic Gene lin-4 Encodes Small RNAs with Antisense Complementarity to lin-14. *Cell* 75, 843–54 (1993).
10. Wightman, B, Ha, I. and Ruvkun, G. Posttranscriptional regulation of the heterochronic gene lin-14 by lin-4 mediates temporal pattern formation in *C. elegans*. *Cell* 75, 855–62 (1993).
11. Reinhart, BJ, Slack, FJ and Basson, M. The 21-nucleotide let-7 RNA regulates developmental timing in *Caenorhabditis elegans*. *Nature* 403, 901–6 (2000).
12. Flynt, AS and Lai, EC. Biological principles of microRNA-mediated regulation: shared themes amid diversity. *Nature Reviews Genetics* 9, 831–42 (2008).
13. Pedersen, I and David, M. MicroRNAs in the immune response. *Cytokine* 43, 391–4 (2008).
14. Sunkar, R, Chinnusamy, V, Zhu, J and Zhu, J-K. Small RNAs as big players in plant abiotic stress responses and nutrient deprivation. *Trends Plant Science* 12, 301–9 (2007).
15. Voinnet, O. Post-transcriptional RNA silencing in plant-microbe interactions: a touch of robustness and versatility. *Current Opinion in Plant Biology* 11, 464–70 (2008).
16. Huijser, P and Schmid, M. The control of developmental phase transitions in plants. *Development* 138, 4117–29 (2011).
17. Rubio-Somoza, I and Weigel, D. MicroRNA networks and developmental plasticity in plants. *Trends in Plant Science* 16, 258–64 (2011).
18. Khan, GA, Declerck, M, Sorin, C, Hartmann, C, Crespi, M and Lelandais-Brière, C. MicroRNAs as regulators of root development and architecture. *Plant Molecular Biology* 77, 47–58 (2011).
19. Nonogaki, H. MicroRNA gene regulation cascades during early stages of plant development. *Plant and Cell Physiology* 51, 1840–6 (2010).
20. Wang, J-W, Park, M-Y, Wang, L-J, Koo, Y, Chen, X-Y, et al. miRNA control of vegetative phase change in trees. *PLoS Genetics* 7, e1002012 (2011).
21. Khraiwesh, B, Zhu, J-K and Zhu, J. Role of miRNAs and siRNAs in biotic and abiotic stress responses of plants. *Biochimica Biophysica Acta* 1819, 137–48 (2012).
22. Lin, H-J, Zhang, Z-M, Shen, Y-O, Gao, S-B and Pan, G-T. Review of plant miRNAs in environmental stressed conditions. *Research Journal Agriculture Biological Sciences* 5, 803–14 (2009).
23. Sunkar, R. MicroRNAs with macro-effects on plant stress responses. *Seminars Cell and Developmental Biology* 21, 805–11 (2010).
24. Xie, Z, Allen, E, Fahlgren, N, Calamar, A, Givan, SA and Carrington, JC. Expression of Arabidopsis MIRNA genes. *Plant Physiology* 138, 2145–54 (2005).
25. Zhou, X, Ruan, J, Wang, G and Zhang, W. Characterization and identification of microRNA core promoters in four model species. *PLoS Computational Biology* 3, e37 (2007).

26. Colaiacovo, M, Lamontanara, A, Bernardo, L, Alberici, R, Crosatti, C, et al. On the complexity of miRNA-mediated regulation in plants: novel insights into the genomic organization of plant miRNAs. *Biology Direct* 7, 15 (2012).
27. Sunkar, R and Zhu, J-K. Novel and stress-regulated microRNAs and other small RNAs from Arabidopsis. *Plant Cell* 16, 2001–19 (2004).
28. Rajagopalan, R, Vaucheret, H, Trejo, J and Bartel, DP. A diverse and evolutionarily fluid set of microRNAs in Arabidopsis thaliana. *Genes and Development* 20, 3407–25 (2006).
29. Zhu, Q-H, Spriggs, A, Matthew, L, Fan, L, Kennedy, G, et al. A diverse set of microRNAs and microRNA-like small RNAs in developing rice grains. *Genome Research* 18, 1456–65 (2008).
30. Li, T, Li, H, Zhang, Y-X and Liu, J-Y. Identification and analysis of seven H_2O_2-responsive miRNAs and 32 new miRNAs in the seedlings of rice (*Oryza sativa* L. ssp. *indica*). *Nucleic Acids Research* 39, 2821–33 (2011).
31. Merchan, F, Boualem, A, Crespi, M and Frugier, F. Plant polycistronic precursors containing non-homologous microRNAs target transcripts encoding functionally related proteins. *Genome Biology* 10, R136 (2009).
32. Gregory, BD, O'Malley, RC, Lister, R, Urich, MA, Tonti-Filippini, J, et al. A link between RNA metabolism and silencing affecting Arabidopsis development. *Developmental Cell* 14, 854–66 (2008).
33. Kim, S, Yang, J-Y, Xu, J, Jang, I-C, Prigge, MJ and Chua, N-H. Two cap-binding proteins CBP20 and CBP80 are involved in processing primary MicroRNAs. *Plant and Cell Physiology* 49, 1634–44 (2008).
34. Laubinger, S, Sachsenberg, T, Zeller, G, Busch, W, Lohmann, JU, et al. Dual roles of the nuclear cap-binding complex and SERRATE in pre-mRNA splicing and microRNA processing in Arabidopsis thaliana. *Proceedings of the National Academy of Sciences of the USA* 105, 8795–800 (2008).
35. Yang, L, Liu, Z, Lu, F, Dong, A and Huang, H. SERRATE is a novel nuclear regulator in primary microRNA processing in Arabidopsis. *The Plant Journal* 47, 841–50 (2006).
36. MacRae, IJ, Zhou, K, Li, F, Repic, A, Brooks, AN, et al. Structural basis for double-stranded RNA processing by Dicer. *Science* 311, 195–8 (2006).
37. Kurihara, Y, Takashi, Y and Watanabe, Y. The interaction between DCL1 and HYL1 is important for efficient and precise processing of pri-miRNA in plant microRNA biogenesis. *RNA* 12, 206–12 (2006).
38. Lobbes, D, Rallapalli, G, Schmidt, DD, Martin, C and Clarke, J. SERRATE: a new player on the plant microRNA scene. *EMBO Reports* 7, 1052–8 (2006).
39. Huang, Y, Ji, L, Huang, Q, Vassylyev, DG, Chen, X and Ma, J-B. Structural insights into mechanisms of the small RNA methyltransferase HEN1. *Nature* 461, 823–7 (2009).
40. Eamens, AL, Smith, NA, Curtin, SJ, Wang, M-B and Waterhouse, PM. The Arabidopsis thaliana double-stranded RNA binding protein DRB1 directs guide strand selection from microRNA duplexes. *RNA* 15, 2219–35 (2009).
41. Mi, S, Cai, T, Hu, Y, Chen, Y, Hodges, E, et al. Sorting of small RNAs into Arabidopsis argonaute complexes is directed by the 5' terminal nucleotide. *Cell* 133, 116–27 (2008).

42. Takeda, A, Iwasaki, S, Watanabe, T, Utsumi, M and Watanabe, Y. The mechanism selecting the guide strand from small RNA duplexes is different among argonaute proteins. *Plant and Cell Physiology* 49, 493–500 (2008).
43. Vaucheret, H. Plant ARGONAUTES. *Trends in Plant Science* 13, 350–8 (2008).
44. Brodersen, P, Sakvarelidze-Achard, L, Bruun-Rasmussen, M, Dunoyer, P, Yamamoto, YY, et al. Widespread translational inhibition by plant miRNAs and siRNAs. *Science* 320, 1185–90 (2008).
45. Axtell, MJ, Jan, C, Rajagopalan, R and Bartel, DP. A two-hit trigger for siRNA biogenesis in plants. *Cell* 127, 565–77 (2006).
46. Allen, E, Xie, Z, Gustafson, AM and Carrington, JC. microRNA-directed phasing during trans-acting siRNA biogenesis in plants. *Cell* 121, 207–21 (2005).
47. Montgomery, TA, Howell, MD, Cuperus, JT, Li, D, Hansen, JE, et al. Specificity of ARGONAUTE7-miR390 interaction and dual functionality in TAS3 trans-acting siRNA formation. *Cell* 133, 128–41 (2008).
48. Law, JA and Jacobsen, SE. Establishing, maintaining and modifying DNA methylation patterns in plants and animals. *Nature Reviews Genetics* 11, 204–20 (2010).
49. Carlsbecker, A, Lee, J-Y, Roberts, CJ, Dettmer, J, Lehesranta, S, et al. Cell signalling by microRNA165/6 directs gene dose-dependent root cell fate. *Nature* 465, 316–21 (2010).
50. Fahlgren, N, Howell, MD, Kasschau, KD, Chapman, EJ, Sullivan, CM, et al. High-throughput sequencing of Arabidopsis microRNAs: evidence for frequent birth and death of MiRNA genes. *PloS One* 2, e219 (2007).
51. Sunkar, R, Li, Y-F and Jagadeeswaran, G. Functions of microRNAs in plant stress responses. *Trends Plant Science* 17, 196–203 (2012).
52. Garcia, D. A miRacle in plant development: role of microRNAs in cell differentiation and patterning. *Seminars Cell and Developmental Biology* 19, 586–95 (2008).
53. Franco-Zorrilla, JM, Valli, A, Todesco, M, Mateos, I, Puga, MI, et al. Target mimicry provides a new mechanism for regulation of microRNA activity. *Nature Genetics* 39, 1033–7 (2007).
54. Sunkar, R, Kapoor, A and Zhu, J-K. Posttranscriptional induction of two Cu/Zn superoxide dismutase genes in Arabidopsis is mediated by downregulation of miR398 and important for oxidative stress tolerance. *Plant Cell* 18, 2051–65 (2006).
55. Kuo, H-F and Chiou, T-J. The role of microRNAs in phosphorus deficiency signaling. *Plant Physiology* 156, 1016–24 (2011).
56. Fujii, H, Chiou, T-J, Lin, S-I, Aung, K and Zhu, J-K. A miRNA involved in phosphate-starvation response in Arabidopsis. *Current Biology* 15, 2038–43 (2005).
57. Pant, BD, Buhtz, A, Kehr, J and Scheible, W-R. MicroRNA399 is a long-distance signal for the regulation of plant phosphate homeostasis. *The Plant Journal* 53, 731–8 (2008).
58. Buhtz, A, Pieritz, J, Springer, F and Kehr, J. Phloem small RNAs, nutrient stress responses, and systemic mobility. *BMC Plant Biology* 10, 64 (2010).

59. Chuck, G and O'Connor, D. Small RNAs going the distance during plant development. *Current Opinion in Plant Biology* 13, 40–5 (2010).
60. Shukla, LI, Chinnusamy, V and Sunkar, R. The role of microRNAs and other endogenous small RNAs in plant stress responses. *Biochimica Biophysica Acta* 1779, 743–8 (2008).
61. Warthmann, N, Chen, H, Ossowski, S, Weigel, D and Hervé, P. Highly specific gene silencing by artificial miRNAs in rice. *PloS One* 3, e1829 (2008).
62. Molesini, B, Pii, Y and Pandolfini, T. Fruit improvement using intragenesis and artificial microRNA. *Trends Biotechnology* 30, 80–8 (2012).
63. Liu, Q and Chen, Y-Q. A new mechanism in plant engineering: the potential roles of microRNAs in molecular breeding for crop improvement. *Biotechnology Advances* 28, 301–7 (2010).
64. Tang, G, Xiang, Y, Kang, Z, Mendu, V, Tang, X, et al. Small RNA technologies: siRNA, miRNA, antagomiR, target mimicry, miRNA sponge and miRNA profiling. In Ying, SY (Ed.) *Current Perspectives in microRNAs*. Springer, Dordrecht, the Netherlands (2008). pp. 17–33. ISBN: 978-1-4020-8533-8.
65. Rubio-Somoza, I, Weigel, D, Franco-Zorilla, J-M, García, JA and Paz-Ares, J. ceRNAs: miRNA target mimic mimics. *Cell* 147, 1431–2 (2011).
66. Todesco, M, Rubio-Somoza, I, Paz-Ares, J and Weigel, D. A collection of target mimics for comprehensive analysis of microRNA function in *Arabidopsis thaliana*. *PLoS Genetics* 6, e1001031 (2010).
67. Rubio-Somoza, I and Manavella, PA. Mimicry technology: suppressing small RNA activity in plants. *Methods in Molecular Biology* 732, 131–7 (2011).
68. Kluiver, J, Gibcus, JH, Hettinga, C, Adema, A, Richter, MKS, et al. Rapid generation of microRNA sponges for microRNA inhibition. *PloS One* 7, e29275 (2012).

3

Epigenetic control by plant Polycomb proteins: new perspectives and emerging roles in stress response

Filomena De Lucia, Institut Pasteur, France, and Valérie Gaudin, Institut Jean-Pierre Bourgin, France

DOI: 10.1533/9781908818478.31

Abstract: In eukaryotes, cell-fate determination, differentiation and developmental programs require precise spatial and temporal control of gene expression. Polycomb group (PcG) proteins are key transcriptional regulators in these mechanisms. Research over the past decade has demonstrated the significance of the Polycomb system in establishing silent chromatin states and perpetuating them through cell divisions. PcG-mediated gene silencing is a fundamental and conserved mechanism employed by both plants and animals. In plants these proteins maintain cell identity and imprinting and control important developmental transitions, hence having agronomical impacts.

In this chapter, we discuss the impact of Polycomb regulation in controlling epigenetic states in plants. We highlight recent reports on the role of long non-coding RNAs as PcG cofactors and the involvement of PcG complexes in stress responses. Indeed, how plants respond to environmental stresses by modifying expression of the genome is a challenge in plant biology, and unexplored links between stress response and PcG regulation are promising.

Key words: Polycomb group (PcG) proteins, chromatin, epigenetic long non-coding RNAs, histone modification, nuclear organization.

3.1 Introduction

Polycomb group (PcG) proteins are key chromatin regulators present in both plants and animals. They work as multi-protein complexes, cooperatively establish silent chromatin states and are involved in the transmission of epigenetic states. Despite the recent progress in both animal and plant fields, PcG mechanisms of action are still poorly understood, impeding the progress towards a full understanding of eukaryote development. How these key chromatin complexes are recruited to specific target genes still remains one of the major questions. Two important achievements in the field have recently had a profound impact on the understanding of the mechanism of action of the Polycomb system and the regulation of PcG target genes: (i) the discovery of non-coding RNAs (ncRNAs) interacting with PcG proteins (Spitale et al., 2011; Wang and Chang, 2011); and (ii) the interplay between nuclear organization and PcG-mediated gene regulation (Bantignies and Cavalli, 2011).

In the plant model *Arabidopsis thaliana*, PcG complexes have been extensively characterized in relation to plant development (for reviews, see Butenko and Ohad, 2011; Holec and Berger, 2012; Köhler and Hennig, 2010), but they have only recently been reported to play a role in perenniality and stress responses. Here, we focus on these questions and we will integrate the current knowledge in the field at the interface with plant responses under stress conditions.

3.2 Conserved multi-protein complexes with histone post-translational modifying activities

Studies in *Drosophila melanogaster* described at least three main types of complexes with different functions which serve as reference types in both animal and plant species: Polycomb repressive complex 1 (PRC1), PRC2, and the Pleiohomeotic repressive complex (Pho-RC) (reviewed in Margueron and Reinberg, 2011; Schuettengruber and Cavalli, 2009; Simon and Kingston, 2009). Pho-RC contains Pleiohomeotic (Pho), a sequence-specific DNA-binding protein, and the Smc-related protein containing four MBT domains (Sfmbt). This complex remains to be described in plants. While no enzymatic activity has been associated with the Pho-RC complex, the two PRC complexes have histone tail post-translational modifying activities. PRC2 catalyses the trimethylation of the histone H3 lysine 27 residue (H3K27me3) at target loci, while PRC1 can recognize H3K27me3 and establish H2A histone monoubiquitination. Thus, PRC1 functions downstream of PRC2 to maintain the repressive state and mediate gene expression control in various biological processes. However, besides the current hierarchical model, which posits two sequential steps by PRC2 and PRC1, respectively, recent studies in animals suggest that H3K27me3 is not always sufficient or required as a docking site for PRC1 recruitment, and that PRC1 may act in a PRC2-independent manner (Gao et al., 2012; Simon and Kingston, 2009; Sing et al., 2009; Tavares et al., 2012). Besides its catalytic activity, PRC1 also has chromatin compaction activity which can be independent of monoubiquitination; however, both activities impact gene expression regulation (Endoh et al., 2012; Eskeland et al., 2010).

A PRC2-type complex is composed of four subunits, among which Enhancer of Zeste (E(Z)) is a SET (Su(var)-E(z)-Thritorax) domain protein that possesses H3K27 histone methyltransferase (HMT) activity. The E(Z) protein on its own has a null or low activity and requires non-catalytic subunits. The other subunits are Suppressor of Zeste 12 (Su(Z)12), the nucleosome remodelling p55 factor and Extra sex combs (Esc) (Figure 3.1).

Su(Z)12 is a protein containing a C2H2 zinc-finger domain and VEFS, a conserved domain among plant and drosophila homologs (VRN2–EMF2–FIS2–Su(Z)12). The p55/NURF55/CAF1 protein and Esc are two WD40 proteins, the former being involved in interaction with

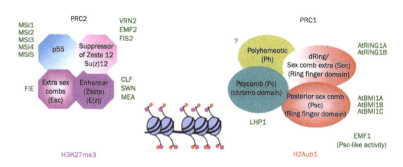

Figure 3.1 Core components of the PRC complexes in Drosophila (in black) and *A. thaliane (in green)*. PRC2 deposits the H3K27me3 silent epigenetic mark whereas H2A monoubiquitination is achieved by PRC1

nucleosomes and the latter is able to modulate the HMT activity. Besides the four core subunits, regulatory subunits (PcG cofactors) have been described in animals that can associate with the PRC2 complex and influence its function, such as Polycomb like (PCL) or JARID2, a histone demethylase essential for developmental control despite impaired enzymatic activity (Landeira and Fisher, 2011).

In *A. thaliana*, extensive genetic analysis of the PcG proteins has been conducted, revealing that PRC2 is structurally conserved and comprises homologs of the four core subunits (reviewed in Köhler and Hennig, 2010). One of the PRC2 core subunits, the Extra sex combs (Esc) homolog, is encoded by a unique gene, FERTILIZATION INDEPENDENT ENDOSPERM (FIE), whereas the three other PRC2 core subunits are encoded by small gene families up to five members. The MULTICOPY SUPPRESSOR OF IRA 1 to 5 genes (MSI1–5) encode the PRC2 p55 homolog, whereas EMBRYONIC FLOWER2 (EMF2), FERTILIZATION INDEPENDENT SEED2 (FIS2) and VERNALIZATION2 (VRN2) are the homologs of the drosophila Su(Z)12 protein. CURLY LEAF (CLF), SWINGER (SWN, also known as EZA1), and MEDEA (MEA) are the homologs of the E(Z) PRC2 histone methyltransferase. The three E(Z) HMTs differ by their expression patterns throughout the plant life cycle. *MEA* is expressed predominantly in the female gametophyte and developing seeds but not in mature seeds and seedlings, whereas *CLF* and *SWN* are present in vegetative tissues. Redundancy issues are certainly responsible for the difficulty of an unambiguous identification of complex composition, and yet scant biochemical data exist on the characterization of the specific complexes. Three main PRC2 complexes have been characterized (Hennig and Derkacheva, 2009; Köhler and Hennig, 2010) and associated with different plant developmental stages (the EMBRYONIC FLOWER (EMF), VERNALIZATION (VRN) and FERTILIZATION INDEPENDENT SEED (FIS) complexes). PcG cofactors similar to PCL have been described in plants. The H3K27 trimethylation activity of the VRN-PRC2 complex is increased by the interaction with three related plant homeodomain (PHD) finger proteins, VEL1, VRN5 and VIN3, that form the PHD-PCR2 complex involved in the vernalization process (De Lucia et al., 2008) (Figure 3.2).

While PRC2 complexes are evolutionarily conserved, PRC1 complexes are more plastic and heterogeneous in eukaryotes, suggesting that they form a small family (Schuettengruber and Cavalli, 2009). The four core components of the drosophila PRC1 complex are: Polycomb (Pc), a protein whose chromodomain recognizes the silencing mark H3K27me3, Polyhomeotic (Ph) (a protein containing SAM and C2-C2 zinc finger

Epigenetic control by plant Polycomb proteins

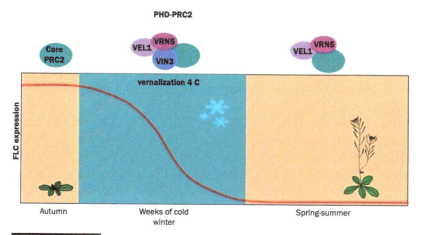

Figure 3.2   The vernalization process and the PRC2 complexes

domains), and two PRC1 RING proteins, drosophila Ring finger protein 1/Sex combs extra (dRing/Sce) and Posterior sex combs (Psc).

The PRC1 RING proteins contain a Ring-finger domain at the N-terminal region and the RAWUL (Ring-finger and WD40 associated Ubiquitin-Like) domain at the C-terminal region, and they are further subdivided into the RING subfamily (dRing/RING1A/RING1B) and the BMI subfamily (Psc/Polycomb group RING fingers (PCGFs)/BMI proteins) (Sanchez-Pulido et al., 2008).

Some of the PRC1 subunits are encoded by small families in mammals and plants, giving rise to a diversity of PRC1 complexes (Figure 3.1). In an extensive proteomic study, six mammalian PRC1 complexes have been reported, corresponding to combinatorial associations of the different PcG homologs, all of which contain RING1A/B and are distinguishable by their Psc homologs (PCGF1 to 6) (Gao et al., 2012). By analysing the genomic localization of different subunits of the six PRC1 complexes, regions bound exclusively or predominantly by one of the six PCGF were identified, highlighting that the different complexes have distinct genomic localizations (Gao et al., 2012). Additional components are also found associated with PRC1 in substoichiometric amounts, such as the PcG protein Sex Comb on Midleg (Scm), and an atypical PRC1 complex, dRING associated Factors (dRAF), comprising dRING, Psc, and the H3K36me2 histone demethylase KDM2 has been reported (Lagarou et al., 2008). Interestingly, the Psc subunit is crucial for inheritance of chromatin features throughout DNA replication (Lo et al., 2012).

In plants, recent data indicate that the chromodomain protein LIKE HETEROCHROMATIN PROTEIN1 (LHP1) (Gaudin et al., 2001) and homologs of the two drosophila RING-finger domain proteins form a complex similar in domain composition and function to animal PRC1 (Bratzel et al., 2010; Chen et al., 2010; Xu and Shen, 2008; Zhang et al., 2007b). The chromodomain of LHP1 binds to trimethylated lysines H3K27 and H3K9me3 *in vitro* (Zhang et al., 2007b). However, LHP1 is mainly localized in euchromatin *in vivo* (Libault et al., 2005), and its genome-wide distribution indicated that LHP1 mainly co-localizes with H3K27me3 in vivo, suggesting that LHP1 is a functional homolog of Pc (Turck et al., 2007; Zhang et al., 2007b). Recent works have identified 5 RING-finger domain proteins interacting with LHP1, AtRING1A/B and AtBMI1A/B/C, AtBMI1A/B having a H2A monoubiquitination activity (Bratzel et al., 2010; Chen et al., 2010; Xu and Shen, 2008). The plant-specific EMF1 protein interacts physically with some of the plant PRC1 subunits, thus participating in a PRC1-like complex (Calonje et al., 2008). Recently, EMF1 was shown to bind tightly to free DNA and to inhibit chromatin remodeling, having a Psc-like activity (Beh et al., 2012). Beside the chromodomain, LHP1 possesses a chromoshadow domain with protein–protein interaction properties (Gaudin et al., 2001). Consistently, several LHP1 partners have been found, such as transcription factors (SCARECROW (Cui and Benfey, 2009) and SHORT VEGETATIVE PHASE (SVP) (Liu et al., 2009)), as well as various chromatin-associated proteins (del Olmo et al., 2010; Li and Luan, 2011; Gaudin, unpublished data). These associated proteins, such as transcription factors, could participate in specifying sets of PcG target genes, as already reported in animals (Yu et al., 2012), giving rise to PRC1 variant complexes, or modulating the plant PRC1 activity.

In the coming years, we expect a rapid development of the biochemical complexity of the plant PRC complexes by the identification of homologous proteins or additional regulatory partners, to fulfil the diverse functions of these complexes in plants. The diversity of the complexes may also be accompanied by variations in the recruitment pathways (histone marks, Polycomb-response elements (PREs), transcription factors, RNA cofactors ...) as well as transcription repression mechanisms (direct interference with the transcription machinery (Lehmann et al., 2012), modulation of chromatin compaction, inhibition of chromatin remodeling, effects on higher-order nuclear structures).

3.3 Polycomb functions in plant development

Plants, which evolved multicellularity independently of animals, utilize the PcG system in strikingly similar ways (Meyerowitz, 2002). The developmental functions of the plant PcG are evident from the multitude of developmental defects exhibited by mutants of various PcG subunit genes (Bratzel et al., 2010; Gaudin et al., 2001; Goodrich et al., 1997; Grossniklaus et al., 1998; Hennig et al., 2003; Luo et al., 1999; Ohad et al., 1999; Xu and Shen, 2008), indicating that PcG complexes play a crucial role in all major stages of plant development and progressions through the plant life cycle. A complete loss of PcG function in plant cells inhibits cell differentiation and activates an embryo-like state (Bratzel et al., 2010; Chanvivattana et al., 2004; Chen et al., 2010), suggesting that PcG is required to control cell identity and the ability to form different tissues with diverse functions. PcG function is also required to maintain gametophytic identity of cells (for a review, see Köhler and Hennig, 2010). Moreover, the system controls one of the most important developmental transitions plants make, the floral transition, from vegetative growth to reproductive flowers. The PcG system regulates flowering through a process called vernalization that allows plants to flower after long periods of cold in more favorable conditions in Spring (Kim et al., 2009). This is achieved via down-regulation of *FLOWERING LOCUS C (FLC)*, whose repression is maintained by the Polycomb system (Bastow et al., 2004; De Lucia et al., 2008; Gendall et al., 2001; Greb et al., 2007; Mylne et al., 2006); for a review, see Kim and Sung (2012) (Figure 3.2).

With the identification of PcG targets at the genomic scale via the analysis of the trimethylated H3K27, signature of Polycomb repression, and the in-depth studies on a few PcG targets in plants, the crucial role of the PcG in plant development is highly appreciated. In *A. thaliana*, H3K27me3 is found in approximately 20% of annotated genes (Zhang et al., 2007a), essentially transcriptional regulators but also genes encoding microRNAs, and fluctuates according to tissues and developmental stages, thus allowing fine-tuning of gene expression during plant differentiation (Lafos et al., 2011). The genome-wide distribution of the PRC1 subunit LHP1 by DamID approach (Germann and Gaudin, 2011) identified about 2300 target regions and 4000 genes distributed over the five chromosomes, of which 90% are associated with H3K27me3 (Zhang et al., 2007). These data confirm the hierarchical mechanism of PRC complexes, but also suggest that PRC1 might act independently of PRC2 for a tiny set of target regions devoid of H3K27me3 (Zhang et al., 2007b).

3.4 Non-coding RNAs as regulatory cofactors of Polycomb complexes

Recent high throughput sequencing approaches have listed thousands of non-coding RNAs (ncRNAs) in mammals with an extraordinary diversity, these molecules being transcribed either from intergenic regions or from intragenic regions in a sense or in antisense orientation. As their functionalities were questioned, the last few years have seen mounting evidence of regulatory roles of ncRNAs, ranging from 50 to several kbp in length, as signals, decoys, guides and/or scaffolds in assembling and regulating various protein complexes and thus impacting gene expression (Wang and Chang, 2011). In particular, their association with chromatin-modifying enzymes and PRC2 complexes has recently been reported (for reviews, see Bracken and Helin, 2009; De Lucia and Dean, 2011; Guttman and Rinn, 2012; Nagano and Fraser, 2011). This emerging theme already has many examples in the animal fields. The pioneer HOTAIR lncRNA regulating the HOX-D locus (Rinn et al., 2007; Tsai et al., 2010), the RepA/Xist RNA involved in X chromosome inactivation (Zhao et al., 2008), and Kcnq1ot1 (Pandey et al., 2008) participating in repression of imprinted genes have all been implicated in Polycomb control. Actually, a significant part of the long interspersed non-coding RNAs (lncRNAs) in the human genome seems to be physically associated with PRC2 (Guttman and Rinn, 2012; Khalil et al., 2009). Control of PRC1 may also involve ncRNA (Yap et al., 2010). Recent work has also identified several shorter RNAs (50–200 nt) as candidate PRC2 regulators, which are transcribed from the 5′ end of PcG target genes (Kanhere et al., 2010). While HOTAIR works in *trans* (Gupta et al., 2010; Rinn et al., 2007), a large number of non-coding transcripts seem to act in *cis* (Figure 3.3). This type of silencing is likely to have widespread consequences, and ncRNAs can now be considered as PcG cofactors that could guide site-specific recruitment of PRC complexes to target genes.

RNA-mediated recruitment is especially attractive for PcG proteins due to the absence of specific DNA-binding components of most PRC complexes. Thus, PcG proteins must rely on other factors that can recognize specific DNA sequences. Drosophila biologists defined small stretches of DNA associated with target genes that are necessary and

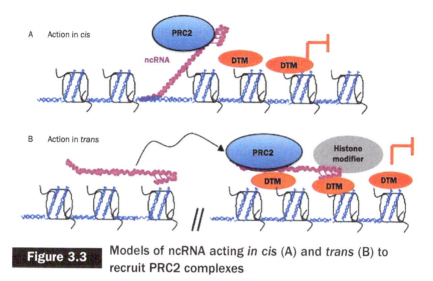

Figure 3.3 Models of ncRNA acting *in cis* (A) and *trans* (B) to recruit PRC2 complexes

sufficient to perform the silencing and maintenance tasks of the PcG system. These sequences have been called Polycomb-response elements (PREs). In flies, Polycomb complexes may contain sequence-specific DNA-binding factors, such as Zeste, Pipsqueak (PSQ) or Polyhomeotic (Pho), to help bind PREs (Ringrose and Paro, 2004; Schwartz and Pirrotta, 2008). By contrast, mammalian and plant Polycomb complexes are not thought to contain such subunits (DNA binding proteins) or PREs, though a few PRE-like elements have recently been reported in mammals (Sing et al., 2009; Woo et al., 2010) and the JARID2 histone demethylase may facilitate binding (Li et al., 2010; Pasini et al., 2010; Peng et al., 2009). Therefore, the mechanism of PcG recruitment to thousands of genomic locations remains poorly understood. In this respect, ncRNAs may be playing a crucial role (Bracken and Helin, 2009; Guenther and Young, 2010). RNA-mediated recruitment represents an ideal mechanism for targeting chromatin modifiers in *cis* to specific alleles or unique locations in the genome, as they remain tethered to the site of transcription and can co-transcriptionally direct enzymatic activities to a unique region (Lee, 2009, 2010). They can also be complementary to genomic sequences and act in *trans*. Besides guiding functions allowing recruitment of chromatin complexes to specific genomic locations, ncRNAs can also act as scaffold elements to assemble large complexes in various processes (Guetg et al., 2012; Wang and Chang, 2011). A recent study has demonstrated the role of HOTAIR in scaffolding PCR2 complex with H3K4 demethylase complex (Tsai et al., 2010). Furthermore, HOTAIR has been shown to interact with thousands of genomic loci, binding very short regions that serve as nucleation sites for PRC silencing (Chu et al., 2011).

To date there is a very limited number of examples of ncRNAs interacting with Polycomb complexes in plants (De Lucia and Dean, 2011; Kim and Sung, 2012; Muller and Goodrich, 2011). Non-coding RNAs have a central role in the regulation of *FLOWERING LOCUS C* (*FLC*), a transcriptional regulator that delays the transition to flowering. *FLC* expression is repressed by prolonged cold experienced during winter (Figure 3.2). This silencing is epigenetically stable and involves a Polycomb-like mechanism and ncRNAs. The ncRNA *COOLAIR* is expressed from a promoter located in the 3′-flanking non-coding region of the *FLC* gene, it is partially spliced and it is involved in the early repression of *FLC* expression during vernalization (Swiezewski et al., 2009). By native RNA immunoprecipitation (RIP), the antisense *COOLAIR* transcript was found interacting with PRC2 (De Lucia et al., unpublished). As *COOLAIR* reaches its peak of expression, the amount of *FLC* mRNA starts to fall and transcription of *COLDAIR* is initiated (Heo and Sung, 2011). *COLDAIR* is a non-coding transcript that links decreased *FLC* expression to the maintenance of its repression by PcG protein complexes in *A. thaliana*. *COLDAIR* is encoded at the *FLC* locus as an intragenic sense RNA located in the first intron of the *FLC* gene. Its expression is increased by cold and it is reported to bind to CLF, the HMT subunit of plant PRC2 complex (Heo and Sung, 2011). However, Heo and Sung show that other proteins are required to ensure specificity of this RNA-protein interaction. Thus it appears that PRC2–lncRNA association is an evolutionarily conserved mechanism in PcG-mediated gene repression in eukaryotes. As for PRC1, a RNA binding protein interacts with LHP1, which may suggest some indirect implication of RNA in gene regulation by plant PRC1-like complex (Latrasse et al., 2011).

From all these data, it seems that non-coding RNAs are at the heart of developmental regulation, determining the epigenetic status and transcriptional network in any given cell type, and they provide a means to integrate external differentiation cues with dynamic nuclear responses.

3.5 Emerging roles of PcG and ncRNAs in responses to environmental stress

The Polycomb system is not only involved in the regulation of plant development. Recent studies show that cold-induced genes (Kwon et al.,

2009) and drought-stress response genes are regulated by PcG (Alexandre et al., 2009; Finnegan et al., 2011). Expression of the genes coding for Polycomb components in barley was found to be responsive to the abiotic stress-related hormone abscisic acid (ABA), implying an association with ABA-mediated processes during seed development and stress responses. Transcriptome analyses highlighted enrichment in stress response genes in *lhp1* mutant (Latrasse et al., 2011).

Besides emerging evidence of the role of PcG proteins in response to environmental stress, ncRNAs may also play a critical role in the adaptation of developmental processes rather than in differentiation *per se*, as very few ncRNAs involved in development were previously revealed by classical genetic approaches. Like some miRNAs, certain ncRNAs are induced in various developmental processes as well as during abiotic stress responses in plants and animals (Jones-Rhoades et al., 2006; Prasanth and Spector, 2007; Sunkar et al., 2007; Yaish et al., 2011). In *A. thaliana*, Crespi's group identified ncRNAs displaying diverse regulation by environmental stimuli (Hirsch et al., 2006). A subsequent study based on genome-wide bioinformatic analysis identified 76 *Arabidopsis* ncRNAs, among which a significant fraction have an altered accumulation in response to abiotic stress (Ben Amor et al., 2009). For some of them the change in expression observed after salt treatment was maintained, while for others the induction was transient. Overexpression of two of these ncRNAs suggested that at least a subset of these newly identified ncRNAs have roles in developmental or stress adaptation programs. The number of examples of ncRNAs responding to stress conditions in plants is definitely increasing (Wu et al., 2012), suggesting that lncRNAs remain an underexplored class of ncRNAs that could be far more prevalent than previously appreciated to adapt growth and development to abiotic interactions. In animals, transcription of lncRNAs within gene regulation elements can modulate activity in response DNA damage (Hung et al., 2011). Hence, lncRNAs represent a class of factors that can be modulated by external stimuli.

3.6 PcG protein functions in three-dimensional nuclear organization

The eukaryotic genome is organized into functionally distinct segments, interspersed and intertwined, that undergo dynamic changes at different phases of the cell cycle and in response to various intracellular and extracellular signals. Furthermore, besides a compartmentalization of the genome into chromosome territories, recent studies using chromosome conformation capture (3C) and derivative techniques have revealed a three-dimensional folding of chromatin into functional globules (Lieberman-Aiden et al., 2009) and an important role of PcG proteins in higher levels of chromatin organization. The three-dimensional organization of Polycomb target genes in the cell nucleus has recently been revealed as an important parameter in their regulation in animals (for a review, see Bantignies and Cavalli, 2011; Delest et al., 2012). Long-distance interactions in *cis* or in *trans* (on different chromosomes) among the major elements bound by PcG proteins, including PREs and core promoters, have been visualized by microscopy-based techniques, 3C and derivative techniques (Comet et al., 2011; Sexton et al., 2012; Tolhuis et al., 2011). Importantly, these interactions change upon activation of a certain target gene, and they can be implicated in the co-regulation of different target genes by PcG proteins via association in three-dimensional nuclear space (Bantignies et al., 2011). Thus, specific nuclear organization imposed by PcG proteins influences the maintenance of epigenetic states and regulates the co-repression of target genes. Among these kinds of contact, 'gene kissing' (between copies of a certain allele) also occurs in *FLC* silencing (De Lucia et al., unpublished), suggesting that they are a general mechanism of PcG regulation in eukaryotes. Thus, besides the classical linear epigenetic organization of the genome, the 3D nuclear architecture has to be taken into account to better understand the complex mechanisms that fine-tune the genetic element activities in various processes and in response to external stimuli.

3.7 Perspectives: the role of Polycomb in abiotic and biotic stress response

Stress puts cells at risk, and adaptation is crucial for maximizing cell survival. Cellular adaptation mechanisms include modification of certain aspects of cell physiology, such as the induction of efficient changes in gene expression programs by intracellular signalling networks. Recent studies using genome-wide approaches as well as single-cell transcription measurements, in combination with classical genetics, have shown that specific modulation of gene expression can be accomplished by several different strategies. How organisms can achieve generic and specific responses to different stresses by regulating gene expression at multiple stages of mRNA biogenesis, from chromatin structure to transcription, mRNA stability and translation, is nowadays a central question.

In the light of the recent evidence concerning the involvement of the Polycomb system and ncRNAs in the abiotic stress response, the identification of Polycomb target genes and the Polycomb RNA partners under different stress conditions is an interesting area to explore. In addition, it has to be taken into account that changes in nuclear organization under stress conditions can allow coordinated co-regulation of a number of genes that need to be modulated in response to stress. Therefore, the recent progress in the identification of a 3D interactome for Polycomb target genes will be crucial for further investigation.

At the same time, the role of chromatin in the memory of the immune response is currently under intense investigation. Priming is a phenomenon that enables cells to respond more rapidly and/or sensitively during a second exposure to a stimulus than during the initial exposure. It is likely that a role for the Polycomb system will be revealed in the near future, in the light of the presence of H3K27me3 in many defense-related genes in Arabidopsis, exploring a Polycomb-mediated chromatin memory for biotic stress responses.

3.8 References

Alexandre C, Moller-Steinbach Y, Schonrock N, Gruissem W, Hennig L (2009) Arabidopsis MSI1 is required for negative regulation of the response to drought stress. *Mol. Plant* 2: 675–87.

Bantignies F, Cavalli G (2011) Polycomb group proteins: repression in 3D. *Trends Genetics* 27: 454–64.

Bantignies F, Roure V, Comet I, Leblanc B, Schuettengruber B, et al. (2011) Polycomb-dependent regulatory contacts between distant Hox loci in Drosophila. *Cell* 144: 214–26.

Bastow R, Mylne JS, Lister C, Lippman Z, Martienssen RA, et al. (2004) Vernalization requires epigenetic silencing of FLC by histone methylation. *Nature* 427: 164–7.

Beh LY, Colwell LJ, Francis NJ (2012) A core subunit of Polycomb repressive complex 1 is broadly conserved in function but not primary sequence. *Proc. Natl. Acad. Sci. U.S.A.* 109: E1063–71.

Ben Amor B, Wirth S, Merchan F, Laporte P, d'Aubenton-Carafa Y, et al. (2009) Novel long non-protein coding RNAs involved in Arabidopsis differentiation and stress responses. *Genome Res.* 19: 57–69.

Bracken AP, Helin K (2009) Polycomb group proteins: navigators of lineage pathways led astray in cancer. *Nature Reviews Cancer* 9: 773–84.

Bratzel F, Lopez-Torrejon G, Koch M, Del Pozo JC, Calonje M (2010) Keeping cell identity in Arabidopsis requires PRC1 RING-finger homologs that catalyze H2A monoubiquitination. *Curr. Biol.* 20: 1853–9.

Butenko Y, Ohad N (2011) Polycomb-group mediated epigenetic mechanisms through plant evolution. *Biochim. Biophys. Acta* 1809: 395–406.

Calonje M, Sanchez R, Chen L, Sung ZR (2008) EMBRYONIC FLOWER1 participates in polycomb group-mediated AG gene silencing in Arabidopsis. *Plant Cell* 20: 277–91.

Chanvivattana Y, Bishopp A, Schubert D, Stock C, Moon YH, et al. (2004) Interaction of Polycomb-group proteins controlling flowering in Arabidopsis. *Development* 131: 5263–76.

Chen D, Molitor A, Liu C, Shen WH (2010) The Arabidopsis PRC1-like ring-finger proteins are necessary for repression of embryonic traits during vegetative growth. *Cell Res.* 20: 1332–44.

Chu C, Qu K, Zhong FL, Artandi SE, Chang HY (2011) Genomic maps of long noncoding RNA occupancy reveal principles of RNA-chromatin interactions. *Molecular Cell* 44: 667–78.

Comet I, Schuettengruber B, Sexton T, Cavalli G (2011) A chromatin insulator driving three-dimensional Polycomb response element (PRE) contacts and Polycomb association with the chromatin fiber. *Proc. Natl. Acad. Sci. U.S.A.* 108: 2294–9.

Cui H, Benfey PN (2009) Interplay between SCARECROW, GA and LIKE HETEROCHROMATIN PROTEIN 1 in ground tissue patterning in the Arabidopsis root. *Plant J.* 58: 1016–27.

De Lucia F, Dean C (2011) Long non-coding RNAs and chromatin regulation. *Curr. Opin. Plant Biol.* 14: 168–73.

De Lucia F, Crevillen P, Jones AM, Greb T, Dean C (2008) A PHD-Polycomb Repressive Complex 2 triggers the epigenetic silencing of FLC during vernalization. *Proc. Natl. Acad. Sci. U.S.A.* 105: 16831–6.

del Olmo I, Lopez-Gonzalez L, Martin-Trillo MM, Martinez-Zapater JM, Pineiro M, et al. (2010) EARLY IN SHORT DAYS 7 (ESD7) encodes the catalytic subunit of DNA polymerase epsilon and is required for flowering repression through a mechanism involving epigenetic gene silencing. *Plant J.* 61: 623–36.

Delest A, Sexton T, Cavalli G (2012) Polycomb: a paradigm for genome organization from one to three dimensions. *Curr. Opin. Cell Biol.* 24: 405–14.

Endoh M, Endo TA, Endoh T, Isono K, Sharif J, et al. (2012) Histone H2A mono-ubiquitination is a crucial step to mediate PRC1-dependent repression of developmental genes to maintain ES cell identity. *PLoS Genetics* 8: e1002774.

Eskeland R, Freyer E, Leeb M, Wutz A, Bickmore WA (2010) Histone acetylation and the maintenance of chromatin compaction by Polycomb repressive complexes. *Cold Spring Harb Symp Quant Biol* 75: 71–8.

Finnegan E, Bond DM, Buzas DM, Goodrich J, Helliwell CA, et al. (2011) Polycomb proteins regulate the quantitative induction of VERNALIZATION INSENSITIVE 3 in response to low temperatures. *Plant J.* 65: 382–91.

Gao Z, Zhang J, Bonasio R, Strino F, Sawai A, et al. (2012) PCGF homologs, CBX proteins, and RYBP define functionally distinct PRC1 family complexes. *Molecular Cell* 45: 344–56.

Gaudin V, Libault M, Pouteau S, Juul T, Zhao G, et al. (2001) Mutations in LIKE HETEROCHROMATIN PROTEIN 1 affect flowering time and plant architecture in Arabidopsis. *Development* 128: 4847–58.

Gendall AR, Levy YY, Wilson A, Dean C (2001) The VERNALIZATION 2 gene mediates the epigenetic regulation of vernalization in Arabidopsis. *Cell* 107: 525–35.

Germann S, Gaudin V (2011) Mapping in vivo protein-DNA interactions in plants by DamID, a DNA adenine methylation-based method. In L Yuan, SE Perry, eds, *Plant Transcription Factors: Methods and Protocols*, Vol. 754, pp. 307–21. Springer Science Business Media, LLC, New York, USA.

Goodrich J, Puangsomlee P, Martin M, Long D, Meyerowitz EM, et al. (1997) A polycomb-group gene regulates homeotic gene expression in *Arabidopsis*. *Nature* 386: 44–51.

Greb T, Mylne JS, Crevillen P, Geraldo N, An H, et al. (2007) The PHD finger protein VRN5 functions in the epigenetic silencing of Arabidopsis FLC. *Curr. Biol.* 17: 73–8.

Grossniklaus U, Vielle-Calzada JP, Hoeppner MA, Gagliano WB (1998) Maternal control of embryogenesis by *MEDEA*, a Polycomb group gene in *Arabidopsis*. *Science* 280: 446–50.

Guenther MG, Young RA (2010) Transcription: Repressive transcription. *Science* 329: 150–1.

Guetg C, Scheifele F, Rosenthal F, Hottiger MO, Santoro R (2012) Inheritance of silent rDNA chromatin is mediated by PARP1 via noncoding RNA. *Molecular Cell* 45: 790–800.

Gupta RA, Shah N, Wang KC, Kim J, Horlings HM, et al. (2010) Long noncoding RNA HOTAIR reprograms chromatin state to promote cancer metastasis. *Nature* 464: 1071–6.

Guttman M, Rinn JL (2012) Modular regulatory principles of large non-coding RNAs. *Nature* 482: 339–46.

Hennig L, Derkacheva M (2009) Diversity of Polycomb group complexes in plants: same rules, different players? *Trends Genet.* 25: 414–23.

Hennig L, Taranto P, Walser M, Schonrock N, Gruissem W (2003) Arabidopsis MSI1 is required for epigenetic maintenance of reproductive development. *Development* 130: 2555–65.

Heo JB, Sung S (2011) Vernalization-mediated epigenetic silencing by a long intronic noncoding RNA. *Science* 331: 76–9.

Hirsch J, Lefort V, Vankersschaver M, Boualem A, Lucas A, et al. (2006) Characterization of 43 non-protein-coding mRNA genes in Arabidopsis, including the MIR162a-derived transcripts. *Plant Physiol.* 140: 1192–204.

Holec S, Berger F (2012) Polycomb group complexes mediate developmental transitions in plants. *Plant Physiol.* 158: 35–43.

Hung T, Wang Y, Lin MF, Koegel AK, Kotake Y, et al. (2011) Extensive and coordinated transcription of noncoding RNAs within cell-cycle promoters. *Nat. Genet.* 43: 621–9.

Jones-Rhoades MW, Bartel DP, Bartel B (2006) MicroRNAs and their regulatory roles in plants. *Annu. Rev. Plant Biol.* 57: 19–53.

Kanhere A, Viiri K, Araujo CC, Rasaiyaah J, Bouwman RD, et al. (2010) Short RNAs are transcribed from repressed polycomb target genes and interact with polycomb repressive complex-2. *Mol. Cell* 38: 675–88.

Khalil AM, Guttman M, Huarte M, Garber M, Raj A, et al. (2009) Many human large intergenic noncoding RNAs associate with chromatin-modifying complexes and affect gene expression. *Proc. Natl. Acad. Sci. U.S.A.* 106: 11667–72.

Kim DH, Doyle MR, Sung S, Amasino RM (2009) Vernalization: winter and the timing of flowering in plants. *Annu. Rev. Cell Dev. Biol.* 25: 277–99.

Kim ED, Sung S (2012) Long noncoding RNA: unveiling hidden layer of gene regulatory networks. *Trends Plant Sci.* 17: 16–21.

Köhler C, Hennig L (2010) Regulation of cell identity by plant Polycomb and trithorax group proteins. *Curr. Opin. Genet. Dev.* 20: 541–7.

Kwon CS, Lee D, Choi G, Chung WI (2009) Histone occupancy-dependent and -independent removal of H3K27 trimethylation at cold-responsive genes in Arabidopsis. *Plant J.* 60: 112–21.

Lafos M, Kroll P, Hohenstatt ML, Thorpe FL, Clarenz O, et al. (2011) Dynamic regulation of H3K27 trimethylation during Arabidopsis differentiation. *PLoS Genet.* 7: e1002040.

Lagarou A, Mohd-Sarip A, Moshkin YM, Chalkley GE, Bezstarosti K, et al. (2008) dKDM2 couples histone H2A ubiquitylation to histone H3 demethylation during Polycomb group silencing. *Genes Dev.* 22: 2799–810.

Landeira D, Fisher AG (2011) Inactive yet indispensable: the tale of Jarid2. *Trends Cell Biol.* 21: 74–80.

Latrasse D, Germann S, Houba-Herin N, Dubois E, Bui-Prodhomme D, et al. (2011) Control of flowering and cell fate by LIF2, an RNA binding partner of the Polycomb complex component LHP1. *PLoS One* 6: e16592.

Lee JT (2009) Lessons from X-chromosome inactivation: long ncRNA as guides and tethers to the epigenome. *Genes Dev.* 23: 1831–42.

Lee JT (2010) The X as model for RNA's niche in epigenomic regulation. *Cold Spring Harb. Perspect. Biol.* 2: a003749.

Lehmann L, Ferrari R, Vashisht AA, Wohlschlegel JA, Kurdistani SK, et al. (2012) Polycomb Repressive Complex 1 (PRC1) disassembles RNA Polymerase II Preinitiation Complexes. *J. Biol. Chem.* 287: 35784–94.

Li G, Margueron R, Ku M, Chambon P, Bernstein BE, et al. (2010) Jarid2 and PRC2, partners in regulating gene expression. *Genes Dev.* 24: 368–80.

Li H, Luan S (2011) The cyclophilin AtCYP71 interacts with CAF-1 and LHP1 and functions in multiple chromatin remodeling processes. *Mol. Plant* 4: 748–58.

Libault M, Tessadori F, Germann S, Snijder B, Fransz F, et al. (2005) The Arabidopsis LHP1 protein is a component of euchromatin. *Planta* 222: 910–25.

Lieberman-Aiden E, van Berkum NL, Williams L, Imakaev M, Ragoczy T, et al. (2009) Comprehensive mapping of long-range interactions reveals folding principles of the human genome. *Science* 326: 289–93.

Liu C, Xi W, Shen L, Tan C, Yu H (2009) Regulation of floral patterning by flowering time genes. *Dev. Cell* 16: 711–22.

Lo SM, Follmer NE, Lengsfeld BM, Madamba EV, Seong S, et al. (2012) A bridging model for persistence of a Polycomb group protein complex through DNA replication in vitro. *Molecular Cell* 46: 784–96.

Luo M, Bilodeau P, Koltunow A, Dennis ES, Peacock WJ, et al. (1999) Genes controlling fertilization-independent seed development in *Arabidopsis thaliana*. *Proc. Natl. Acad. Sci. U.S.A.* 96: 296–301.

Margueron R, Reinberg D (2011) The Polycomb complex PRC2 and its mark in life. *Nature* 469: 343–9.

Meyerowitz EM (2002) Plants compared to animals: the broadest comparative study of development. *Science* 295: 1482–5.

Muller R, Goodrich J (2011) Sweet memories: epigenetic control in flowering. *F1000 Biol. Rep.* 3: 13.

Mylne JS, Barrett L, Tessadori F, Mesnage S, Johnson L, et al. (2006) LHP1, the Arabidopsis homologue of Heterochromatin Protein1, is required for epigenetic silencing of FLC. *Proc. Natl. Acad. Sci. U.S.A.* 103: 5012–17.

Nagano T, Fraser P (2011) No-nonsense functions for long noncoding RNAs. *Cell* 145: 178–81.

Ohad N, Yadegari R, Margossian L, Hannon M, Michaeli D, et al. (1999) Mutations in *FIE*, a WD polycomb group gene, allow endosperm development without fertilization. *Plant Cell* 11: 407–15.

Pandey RR, Mondal T, Mohammad F, Enroth S, Redrup L, et al. (2008) Kcnq1ot1 antisense noncoding RNA mediates lineage-specific transcriptional silencing through chromatin-level regulation. *Mol. Cell* 32: 232–46.

Pasini D, Cloos PA, Walfridsson J, Olsson L, Bukowski JP, et al. (2010) JARID2 regulates binding of the Polycomb repressive complex 2 to target genes in ES cells. *Nature* 464: 306–10.

Peng JC, Valouev A, Swigut T, Zhang J, Zhao Y, et al. (2009) Jarid2/Jumonji coordinates control of PRC2 enzymatic activity and target gene occupancy in pluripotent cells. *Cell* 139: 1290–302.

Prasanth KV, Spector DL (2007) Eukaryotic regulatory RNAs: an answer to the 'genome complexity' conundrum. *Genes Dev.* 21: 11–42.

Ringrose L, Paro R (2004) Epigenetic regulation of cellular memory by the Polycomb and Trithorax group proteins. *Annu. Rev. Genet.* 38: 413–43.

Rinn JL, Kertesz M, Wang JK, Squazzo SL, Xu X, et al. (2007) Functional demarcation of active and silent chromatin domains in human HOX loci by noncoding RNAs. *Cell* 129: 1311–23.

Sanchez-Pulido L, Devos D, Sung ZR, Calonje M (2008) RAWUL: a new ubiquitin-like domain in PRC1 ring finger proteins that unveils putative plant and worm PRC1 orthologs. *BMC Genomics* 9: 308.

Schuettengruber B, Cavalli G (2009) Recruitment of polycomb group complexes and their role in the dynamic regulation of cell fate choice. *Development* 136: 3531–42.

Schwartz YB, Pirrotta V (2008) Polycomb complexes and epigenetic states. *Curr. Opin. Cell Biol.* 20: 266–73.

Sexton T, Yaffe E, Kenigsberg E, Bantignies F, Leblanc B, et al. (2012) Three-dimensional folding and functional organization principles of the Drosophila genome. *Cell* 148: 458–72.

Simon JA, Kingston RE (2009) Mechanisms of Polycomb gene silencing: knowns and unknowns. *Nat. Rev. Mol. Cell Biol.* 10: 697–708.

Sing A, Pannell D, Karaiskakis A, Sturgeon K, Djabali M, et al. (2009) A vertebrate Polycomb response element governs segmentation of the posterior hindbrain. *Cell* 138: 885–97.

Spitale RC, Tsai MC, Chang HY (2011) RNA templating the epigenome: long noncoding RNAs as molecular scaffolds. *Epigenetics* 6: 539–43.

Sunkar R, Chinnusamy V, Zhu J, Zhu JK (2007) Small RNAs as big players in plant abiotic stress responses and nutrient deprivation. *Trends Plant Sci.* 12: 301–9.

Swiezewski S, Liu F, Magusin A, Dean C (2009) Cold-induced silencing by long antisense transcripts of an Arabidopsis Polycomb target. *Nature* 462: 799–802.

Tavares L, Dimitrova E, Oxley D, Webster J, Poot R, et al. (2012) RYBP-PRC1 complexes mediate H2A ubiquitylation at Polycomb target sites independently of PRC2 and H3K27me3. *Cell* 148: 664–78.

Tolhuis B, Blom M, Kerkhoven RM, Pagie L, Teunissen H, et al. (2011) Interactions among Polycomb domains are guided by chromosome architecture. *PLoS Genet.* 7: e1001343.

Tsai MC, Manor O, Wan Y, Mosammaparast N, Wang JK, et al. (2010) Long noncoding RNA as modular scaffold of histone modification complexes. *Science* 329: 689–93.

Turck F, Roudier F, Farrona S, Martin-Magniette ML, Guillaume E, et al. (2007) Arabidopsis TFL2/LHP1 specifically associates with genes marked by trimethylation of histone H3 lysine 27. *PLoS Genet.* 3: e86.

Wang KC, Chang HY (2011) Molecular mechanisms of long noncoding RNAs. *Molecular Cell* 43: 904–14.

Woo CJ, Kharchenko PV, Daheron L, Park PJ, Kingston RE (2010) A region of the human HOXD cluster that confers Polycomb-group responsiveness. *Cell* 140: 99–110.

Wu J, Okada T, Fukushima T, Tsudzuki T, Sugiura M, et al. (2012) A novel hypoxic stress-responsive long non-coding RNA transcribed by RNA polymerase III in Arabidopsis. *RNA Biology* 9: 302–13.

Xu L, Shen WH (2008) Polycomb silencing of KNOX genes confines shoot stem cell niches in Arabidopsis. *Curr. Biol.* 18: 1966–71.

Yaish MW, Colasanti J, Rothstein SJ (2011) The role of epigenetic processes in controlling flowering time in plants exposed to stress. *J. Exp. Botany* 62: 3727–35.

Yap KL, Li S, Munoz-Cabello AM, Raguz S, Zeng L, et al. (2010) Molecular interplay of the noncoding RNA ANRIL and methylated histone H3 lysine 27 by Polycomb CBX7 in transcriptional silencing of INK4a. *Mol. Cell* 38: 662–74.

Yu M, Mazor T, Huang H, Huang HT, Kathrein KL, et al. (2012) Direct recruitment of Polycomb repressive complex 1 to chromatin by core binding transcription factors. *Mol. Cell* 45: 330–43.

Zhang X, Clarenz O, Cokus S, Bernatavichute YV, Pellegrini M, et al. (2007a) Whole-genome analysis of histone H3 lysine 27 trimethylation in Arabidopsis. *PLoS Biol.* 5: e129.

Zhang X, Germann S, Blus BJ, Khorasanizadeh S, Gaudin V, et al. (2007b) The Arabidopsis LHP1 protein colocalizes with histone H3 Lys27 trimethylation. *Nat. Struct. Mol. Biol.* 14: 869–71.

Zhao J, Sun BK, Erwin JA, Song JJ, Lee JT (2008) Polycomb proteins targeted by a short repeat RNA to the mouse X chromosome. *Science* 322: 750–6.

4

Metabolite profiling for plant research

*Nalini Desai and Danny Alexander,
Metabolon, Inc., USA*

DOI: 10.1533/9781908818478.49

Abstract: Metabolomics is a valuable technology for improving our understanding of the physiology and biochemistry of organisms. The analysis of metabolites produced by microorganisms, cells, and different tissues helps in understanding the biosynthetic and catabolic pathways present and how those pathways are affected by genetic and environmental factors. In plants, this valuable technique is being applied to many aspects of plant biology, including growth and development, responses to external stresses, genetics, and nutritional requirements. Thus, it complements other Omics technologies applied to elucidate the functionality of cells, tissues, and plant systems. Wide adoption of metabolic profiling in plants will eventually create synergistic payoffs in our understanding of metabolic regulation in plant biology.

Key words: metabolite profiling, mass spectrometry (MS), chromatographic separation, gas phase (GC/MS), liquid phase (LC/MS), nuclear magnetic resonance (NMR), principal component analysis, genome wide association studies (GWAS).

4.1 Introduction

Metabolite profiling, or metabolomics, the non-targeted study of global changes in small molecule metabolites, is a valuable technology for improving our understanding of physiology and biochemistry, and can be applied to many aspects of plant biology, including growth and development, responses to external stresses, genetics, and nutritional requirements, among others. The incidence and activity of biochemical pathways represent the phenotypic outcome of the genomic makeup, transcript expression, and protein function (enzyme activity, kinetics, substrate and cofactor availability, subcellular location, etc.) in a given organism at a particular time. The biochemistry and regulation of specific metabolic pathways have been elucidated with elegant experimentation over several centuries, but only in the last decade have new technologies been developed and optimized that enable us to view an organism's metabolic changes in a global, non-targeted manner, affording the opportunity to explore in a new way the many unresolved questions in plant biology.

The successful implementation of metabolomics as a discovery platform entails the generation of consistent and reproducible identification and measurement, in a medium to high throughput format, of a broad and useful range of metabolites in a given sample. It culminates in the generation of experimentally testable biological hypotheses, which can serve as the framework for directed and targeted studies. In this review we will discuss the steps that we have taken to develop and optimize an approach to metabolomics that is general, rapid, and widely available. We will provide examples of its application as a discovery platform in a diverse array of biological systems, with a focus on plant biology. We will also briefly describe the impact of plant metabolomics up to the present, and discuss some applications of plant metabolomics related to crop improvement, some already demonstrated in principle, which may be fruitful in the future.

4.2 Methodological approach

In the last 10 years, mass spectrometry (MS) has become the most prevalent detection method used to identify and quantify small biological molecules, typically following chromatographic separation in the gas phase (GC/MS) or the liquid phase (LC/MS). Its superior sensitivity relative to other methods (e.g. nuclear magnetic resonance (NMR)), combined with its information richness, speed, and robustness, has made MS platforms the preferred technology for metabolomic analysis. The goal of metabolomics is to provide a broad and unbiased view of metabolites in a biological system, but, because biological chemicals include an astounding array of varying chemical properties, no single analytical system can provide complete coverage. Definitions of 'complete coverage' are functional in nature, always dependent on the methodologies employed, and especially on their limits of detection. Considering the diversity of compound properties, any attempt to fully catalogue all compounds would require a very large number of separation and detection methods. Also, the compounds present in an organism comprise many different subsets, depending on the tissue, developmental stage, and environmental conditions. Nevertheless, using a few robust and rapid analytical systems (as described below), highly valuable biological information can be discerned from steady-state semi-quantitative measurements of a tractable number of compounds (i.e. a few hundred) and samples (e.g. a few dozen). The platform's practical value lies in the reproducibility, robustness, and especially the speed at which annotated datasets can be completed. The interpretive value, always dependent on good study design, lies in the broad and unbiased nature of the compounds measured, and the fact that perturbations in almost all the major pathways can be observed.

Because metabolite structures cannot be deduced from genomic information, as is the case for RNA and proteins,[1] accurate platform identification of the metabolites and how they fit into metabolic pathways is essential for meaningful biological interpretation. We have therefore adopted a library-based 'chemo-centric' approach to metabolomic analysis,[2] as opposed to 'ion-centric' methods, which treat all ion features independently and then use statistical filtering for data reduction. A chemical library entry is typically based on the performance of an authentic standard in the analytical platforms, which the computer then uses to perform identity searches in raw data. Also, library entries for compounds of unknown identity ('unnamed' compounds) can be constructed using library-building software which accounts for all ion features associated with the unidentified chemical. This approach allows reliable quantification and statistical analysis in a manner similar to that for named compounds, and ensures that specific unnamed compounds are reproducibly measured over time and across datasets. The library parameters, especially fragmentation patterns and retention times, often give valuable structural information which allows the placement of a compound into a particular chemical class (e.g. flavonoid, oxylipin, etc.).

Given the utility of MS platforms for producing high quality biochemical data, the greatest challenges to their practical application have involved managing the sheer amount of data generated, and the need to complete the analysis of a dataset in a reasonable time span and in a reproducible manner. Each sample yields thousands of ion features, the vast majority of which represent either redundant information or process artifacts. It is therefore critical to automate (in software) rapid and robust methods for sample and data management, peak integration, chemical identification, and final data export.[3]

It is important to keep in mind that metabolomics, while heavily driven by the techniques of chemistry, is fundamentally a study of biology. Therefore, the greatest emphasis should be on the biological understanding of data, and how it may reveal the inner workings of organisms. Great care must be given to the study design in terms of biological relevance, statistical power, and accurate execution.[4-8]

4.3 Metabolomic platform

The metabolomic process can be envisioned as four steps, including:

- sample extraction
- data generation
- data quality control and curation
- statistical analysis and biological interpretation.

Simple methanol:water extraction has proven the most general and practical method for the recovery of a broad range of important biochemicals in primary, intermediary, and secondary metabolism, and we have developed rapid and automated methods for sample extraction and liquid handling, minimizing human error and maximizing throughput. In our system, a single extraction supplies aliquots for three MS platforms, run in parallel. These include: 1) GC/MS; 2) an ultra high performance liquid chromatography (UHPLC)/MS system optimized for positive ionization (LC/MS-POS); and 3) UHPLC/MS optimized for negative ionization (LC/MS-NEG). Detailed methods for these have been published previously.[9-11] Together these three platforms provide a balanced and complementary approach to achieve broad coverage. For instance, the GC/MS is optimal for many compounds, such as carbohydrates, which are difficult to separate by typical reverse phase methods. Approximately 70–80% of the compounds detected are measured in at least two platforms, while 30–40% are measured in all three. However, to ensure that a compound is represented only once in the statistical analysis, the result from only one platform, chosen based on analytical characteristics (best signal-to-noise, least interference, etc.), is exported to the final dataset for statistical analysis. Through correlative comparisons, redundant identification on independent platforms serves to lend additional confidence to a compound's identification.

Data quality is monitored at several levels during data generation.[3] Instrument reproducibility is reported by internal standards which are added to each sample just before injection, and also serve as chromatographic retention standards. These are monitored automatically in real time, and failure to detect a proper standard leads to operator intervention. Overall process variation (the combination of extraction, recovery, and instrument variance) is monitored through the analysis of all biochemicals detected in several technical replicates, run at spaced intervals in each run sequence. These technical replicates are taken from a pool made up of small aliquots from the actual biological samples under study, and thus represent an 'average' pool in the appropriate matrix. This approach also has the advantage of providing a measure of analytical reproducibility without the need to run technical replicates of each experimental sample, resulting in substantial time and cost savings.

Immediately following data collection, a suite of software tools integrates peaks for all ion features, reduces data for incorporation into a relational database, and searches for matches against all library entries.[2] The requirement for a combination of correct chromatographic retention, nominal mass, primary ion spectrum, and fragmentation spectra results in robust chemical identification. The integrated peak ion count for a single ion, typically the largest in the spectrum, is used to represent the relative quantification of that compound in the dataset. While absolute quantitation of compounds is always preferable in biological studies when possible, this is not practical in a generalized discovery platform. We would argue that, as a discovery tool, relative changes within a good study design are highly valuable for biological hypothesis generation, and in our experience the rate of successful subsequent validation of metabolomic measurements is extremely high.

A major bottleneck in the production of metabolomic data, yet representing a most critical process, is the verification of accurate peak integration and chemical identification. Individual visual inspection of spectra for hundreds of compounds in dozens (or more) of samples can take weeks or months by manual methods. Again, innovative software tools are used to provide a solution;[3] for each chemical all the samples within a study can be simultaneously visualized in a graphical display of chromatographic properties, peak intensities, mass, and so on, containing rapid links to underlying peak scans and mass spectra. This enables manual acceptance or rejection of calls from the graphical interface for all samples at once, in most cases requiring only a few seconds or minutes per compound. The complete curation process, requiring as little as one day's time, leads to a finalized non-redundant dataset containing only those compounds which meet high confidence identification criteria.

Many statistical techniques can be used to analyze metabolomic datasets (Principal Component Analysis, pair-wise *t*-tests, Random Forest, analysis of variance (ANOVA), etc.), and it is important to consider up front the best study design to optimize chances of revealing biologically relevant answers. Proper controls must be built into the sample set and all samples must be run in a randomized block design, due to the semi-quantitative nature of metabolomic data. Sufficient statistical power is critical, and by far the most important consideration in this regard is understanding the reproducibility of the phenotype under study (i.e. within-group biological variation). Metabolomic data in well-behaved biological systems (e.g. microbes grown in culture) can reveal subtle but real effects using only a small number of biological replicates. However, noisier systems (e.g. field-grown plants) require more replication to achieve similar results (or are restricted to seeing only larger differences). Provided with informative statistical result outputs, data visualization tools, and biochemical pathway organization, a biologist is well equipped to mine metabolomic data for important biological effects, and thus maximize the chance of developing viable hypotheses about biological mechanisms.

One advantage in metabolomic analysis is the ability to consider metabolic context in statistical interpretations. Metabolites in biological pathways and networks systems often respond in coordinated and interpretable ways which may give support to observations beyond formal statistical values, and may suggest novel hypotheses. Also, in a broader sense, frequent findings of well-documented biological effects in the data (e.g. known oxidative stress markers, disease resistance markers, etc.) lend greater confidence to the analytical validity of novel observations.

4.4 Metabolomics in plant science

In plants GC/MS was used as early as 1991 to study the response of barley seedlings to the effects of herbicides.[12] By the end of the decade use of this technology was more widespread,[13,14] and it was being recognized as a potentially powerful tool to complement genome sequencing, RNA profiling, and other technologies in the field of functional genomics.[15–18] In the following 5 years, metabolomic analysis in plant science expanded rapidly (see a 2006 review by Hall[19]), being applied to physiological problems such as growth and development, responses to nutrient stress, environmental stress, and the relationship of secondary metabolites to biotic stress, among others. The last six years have seen a tremendous growth in the field; a few examples of this include applications to carbon–nitrogen regulation,[20,21] responses to pathogens or pests,[22–25] the physiology of plant growth and development,[26,27] phosphorus regulation,[28] abiotic stress,[29–36] and genetic analysis,[37,38] including mutations or transgenic manipulation.[39–42] For a more comprehensive overview of the field, see the many useful reviews of the technology and its application to plants.[19,34,43–49]

4.5 The future role of metabolomics in crop improvement

Enhanced productivity of crop plants remains the foremost goal of developing new varieties of crops, whether those plants are part of large corn fields in the American Midwest or small subsistence farms on marginal land in Sub-Saharan Africa. In modern agriculture, crop plants such as maize hybrids have been bred for several decades to maximize yield. In these hybrids, higher yield has been achieved by the interaction of enhanced genetics to increase yield potential and improved agronomic practices.[50] Differences in the yield potential between genotypes become obvious only in planting at high density, when nutrients become limiting, that is, the genetic ability of the plant to assimilate more nutrients from the environment, and to respond favorably to stress, is the basis of its increased yield potential.[51] The major agronomic traits targeted for increased yield are nitrogen and water use efficiency, and mitigation of various stresses, both biotic and abiotic.[52,53] The success of agronomic trait development is measured by the increase, or conservation, of optimum yield under conditions of stress. In plant metabolic terms, this means the initial availability of metabolites for robust vegetative growth, followed by a timely transition to floral development and seed set, with preferential channeling of nutrients into the developing seed. Optimum environmental conditions before pollination, including adequate nutrition, have been reported to be crucial for the normal development of the seed.[54]

In the last century great strides were made in understanding the biochemistry and physiology of plant carbon and soil nutrient assimilation, and of stress tolerance.[55–59] Yet, we do not completely understand how the progression and regulation of these processes result in increased yield. Unlike the single-gene traits developed in the first phases of plant biotechnology, the 'second generation' yield-related traits are surely affected by multiple genes in complex regulatory networks, and are therefore much more difficult to predict and manipulate. The great challenge of 'omics' technology for now is incomplete annotation. The functions of transcription factors and other proteins, and thus the transcripts and genes, are typically not apparent from sequence alone, so a large proportion remains mysterious. Metabolite data have less complexity than gene or transcript data, and extensive research over the last century has led to a high fraction of the biochemicals having been mapped onto biochemical pathways (i.e. better annotation). Analysis of transcriptomics and proteomics data can be greatly aided by correlation with metabolomic data, to the extent that biochemical pathways are regulated at the levels of transcription, translation, protein turnover, and so on.[1,60]

As discussed above, functional genomics, the integrated study of biomolecules and their expression at all levels, has now begun to be greatly impacted by metabolite profiling technologies. Metabolomics also has the possibility of playing a valuable role in the development of improved complex traits for crop enhancement.[34,37,60-62] First, global chemo-centric metabolic profiling represents a valuable tool for studying changes in plant metabolic pathway expression at various stages in growth and development, in conditions of environmental and nutrition stresses, or resulting from genetic effects. Second, metabolomics has the potential to enable the development of predictive biochemical biomarker assays to shortcut time-consuming breeding steps and yield trials, and to facilitate the discovery and validation of target molecules that might be manipulated to engineer plants with enhanced agronomic traits. A few examples of these approaches are discussed below.

4.5.1 Nitrogen use efficiency

To realize the yield potential of a crop, farmers typically apply large quantities of petroleum-based synthetic fertilizer. Developing genotypes with enhanced utilization of minerals from fertilizer applications, particularly nitrogen, has become a priority due to environmental problems created by excess nitrogen run-off into rivers and leaching into ground water, as well as the escalating cost of fertilizer. Nitrogen use efficiency (NUE) in crops is defined as the increase in yield per unit nitrogen applied. The various steps of nitrogen assimilation are now well understood and have been the subject of excellent recent reviews.[63–65] There are many stages at which nitrogen assimilation in a plant may be regulated: nitrogen uptake, transport, assimilation into cellular components such as amino acids, nucleic acids and co-factors during vegetative growth, or at the stage of nitrogen remobilization and storage during reproductive growth. The timely transition of nitrogen utilization from optimum vegetative growth to storage effectively leads to increased yield, which translates into increased nitrogen use efficiency.[27] In spite of extensive experimentation, our knowledge of integrated signals that affect the transformation from nitrogen assimilation into vegetative tissue to channeling it into storage in seed tissue is limited at present. Further, carbon and nitrogen metabolism regulate each other in maintaining a fine carbon–nitrogen balance within the plant, and perturbation of either process results in poor outcomes. One approach used in attempts to enhance NUE has been to overexpress key enzymes involved in the nitrogen utilization biochemical pathways. Experiments in which transgenic plants overexpressed nitrogen assimilation enzymes such as glutamine synthetase[59,66] or nitrogen transporters[67] have met with limited success. Indeed, the greatest success to date has been achieved by overexpression of alanine transaminase.[68,69] Based on prevailing knowledge of the nitrogen assimilation process, this result was unexpected. Indeed, the traditional approach of designing experiments depending on knowledge-based hypotheses becomes limiting when many interacting biochemical pathways and development stages are found to be impacted in complex traits such as enhanced NUE. Keeping these challenges in mind, researchers have conducted transcriptomics analysis to obtain a better understanding of the global changes that occur in a plant under nitrogen depleted and nitrogen replete conditions.[63,70,71] Monitoring global changes in biochemical profiles of source, (leaf) transport (phloem) and sink (kernel) tissues at various stages of development in a plant, under several nitrogen regimens, may provide direct information on which pathways may need to be manipulated to increase nitrogen utilization efficiency.

4.5.2 Metabolite association with genotype

Association of genome regions (quantitative trait loci (QTL)) with phenotype has been an important breeding tool in the development of new hybrids. However, mapping of QTLs by the traditional linkage analysis has many limitations.[72,73] Genome wide association studies

(GWAS), which aim to map the correlation of genomic sequence diversity of individuals with phenotypic variations,[74] provide a much higher resolution for dissecting the genetic basis of complex traits and identifying specific genes associated with phenotypes. Facilitated by the availability of high density single nucleotide polymorphism (SNP) maps of the human genome, these studies have been developed and widely used for studying the genetic variations underlying disease. The vast natural genotypic variability in plants makes these methods very powerful for understanding plant physiology, and for the discovery of novel traits. High density SNP maps exist in various crop plants such as rice and maize, and in model systems such as *Arabidopsis thaliana*, and several GWA studies in these plants have recently been published.[38,75–78]

One limitation of GWAS in the studies referenced above is that, while genetic associations are readily observed with the phenotype being tested, often the underlying mechanism responsible for the phenotype is difficult to assess, due to the lack of annotation of many genes. The idea that association of metabolic profiles, as an intermediary phenotype in GWAS, would be a powerful tool to advance the mechanistic understanding of complex traits was first tested by Gieger et al.[73] in human serum, and a more recent study[79] advanced this idea further, by testing the association of SNPs with non-targeted global metabolic profiling of serum, analyzing >250 metabolites from 60 biochemical pathways, in a large cohort of 2820 individuals. Thirty-seven loci with significant associations were identified, including 23 loci which describe new genetic associations with metabolic traits (one of which resulted in the discovery of function for a previously uncharacterized gene), and 14 loci that replicate and extend established disease-related SNP associations. Importantly, the utility of GWA mapping of plant metabolic profiles has recently been demonstrated in maize, a major crop plant.[38] In that study 289 diverse maize inbred lines, having been mapped for >50 000 single SNPs, were subjected to metabolomic analysis. Profiling of 118 metabolites in leaves of young plants showed 26 metabolites with strong SNP associations, including several lignin precursors which mapped to a section of chromosome 9 bearing the gene for cinnamoyl-CoA reductase, a key enzyme in monolignol synthesis.

This technology can clearly be useful in the discovery and validation of targets for the development of improved traits, as demonstrated by the monolignol enzyme cited above. In another example, in which genetic association with drought was studied in maize,[77] two GWAS-identified genes were found to be in gene classes which suggested possible links to drought tolerance. In such a study, inclusion of metabolic profiling as an intermediary phenotype might give additional clues about mechanisms of action for these genes, and enable development of specific markers to develop drought-resistant genotypes through breeding or pathway engineering. Such insights into the mechanism of drought tolerance would almost certainly be translatable to other crop species.

4.5.3 Identification of metabolite biomarkers for complex trait prediction

Successful development of novel biotechnology-based products with enhanced agronomic traits would be greatly aided by (and may require) the availability of medium throughput greenhouse assays predictive of crop yield, for both discovery and validation of leads. In the development of single gene herbicide and insect resistance products, hundreds of transgenic plants were typically evaluated in the greenhouse for expression and activity, in order to obtain a few that were good enough for field testing, even though the discovery and validation of the target genes were performed in assays independent of the target plant. The development of yield prediction assays would entail the identification and validation of biochemicals that, in a rapid screen (e.g. a greenhouse seedling test), correlate with yield in actual field situations. Using more than one biochemical marker, particularly a combination of positively and negatively correlating biochemicals, may result in assays with stronger predictability. Therefore, development of metabolite biomarkers that predict complex traits like yield would greatly enhance the rate of agronomic trait development, whether through transgenic or traditional breeding approaches. Predictive markers have long been used in the medical sciences (e.g. glucose, cholesterol) and, while most are now understood in mechanistic detail, this is not an absolute requirement. Many other metabolite biomarkers are currently being developed to assess the risk and progression of various diseases, from cancer to periodontal disease.[80–86]

4.6 Conclusion

The utility of metabolomic analysis in plant biology has by now been well demonstrated, yielding valuable insights into mechanisms of biochemical regulation of growth, development, nutrient and environmental stress, and gene function, among others. However, these represent only a tiny fraction of the potential. As the technology itself advances, we should expect even greater information content and improved biological insights. The challenge is to make this technically difficult technology accessible to the entire plant biology community in a manner that yields high quality, consistent, comparable, and timely results. Wide adoption of metabolic profiling in plants will eventually create synergistic payoffs in our understanding of metabolic regulation in plant biology.

4.7 References

1. Fernie, AR and Stitt, M. On the discordance of metabolomics with proteomics and transcriptomics: coping with increasing complexity in logic, chemistry and network interactions. *Plant Physiology* **158**, 1139–45 (2012).
2. Dehaven, CD, Evans, AM, Dai, H and Lawton, KA. Organization of GC/MS and LC/MS metabolomics data into chemical libraries. *Journal of Cheminformatics* **2**, 9 (2010).
3. DeHaven, CD, Evans, A, Dai, H and Lawton, KA. Software techniques for enabling high-throughput analysis of metabolomic datasets. In U Roessner (Ed.), *Metabolomics* (Vienna: InTech, 2012).
4. Biais, B, Bernillon, S, Deborde, C, Cabasson, C, Rolin, D, et al. Precautions for harvest, sampling, storage, and transport of crop plant metabolomics samples. *Methods in Molecular Biology* **860**, 51 (2012).
5. Fernie, AR, Aharoni, A, Willmitzer, L, Stitt, M, Tohge, T, et al. Recommendations for reporting metabolite data. *The Plant Cell* **23**, 2477–82 (2011).
6. Kim, HK and Verpoorte, R. Sample preparation for plant metabolomics. *Phytochemical Analysis* **21**, 4–13 (2010).
7. Poorter, H, Fiorani, F, Stitt, M, Schurr, U, Finck, A, et al. The art of growing plants for experimental purposes: a practical guide for the plant biologist. *Functional Plant Biology* http://dx.doi.org/10.1071/FP12028 (2012).
8. Tohge, T, Mettler, T, Arrivault, S, Carroll, AJ, Stitt, M and Fernie, AR. Frontiers: From models to crop species: caveats and solutions for translational metabolomics. *Frontiers in Plant Physiology* **2**, 61 (2011).
9. Evans, AM, DeHaven, CD, Barrett, T, Mitchell, M and Milgram, E. Integrated, nontargeted ultrahigh performance liquid chromatography/electrospray ionization tandem mass spectrometry platform for the identification and relative quantification of the small-molecule complement of biological systems. *Analytical Chemistry* **81**, 6656–67 (2009).
10. Lawton, KA, Berger, A, Mitchell, M, Milgram, KE, Evans, AM, et al. Analysis of the adult human plasma metabolome. *Pharmacogenomics* **9**, 383–97 (2008).
11. Ohta, T, Masutomi, N, Tsutsui, N, Sakairi, T, Mitchell, M, et al. Untargeted metabolomic profiling as an evaluative tool of fenofibrate-induced toxicology in Fischer 344 male rats. *Toxicologic Pathology* **37**, 521–35 (2009).

12. Sauter, H, Lauer, M and Fritsch, H. Metabolic profiling of plants—a new diagnostic technique. (pp. 288–99). In DR Baker, JG Fenyes, and MK Moberg (Eds), *Synthesis and Chemistry of Agrochemicals II*. (ACS Symposium Series 443, Washington, DC: American Chemical Society 1991).
13. Adams, MA, Chen, ZL, Landman, P and Colmer, TD. Simultaneous determination by capillary gas chromatography of organic acids, sugars, and sugar alcohols in plant tissue extracts as their trimethylsilyl derivatives. *Analytical Biochemistry* 266, 77–84 (1999).
14. Katona, ZF, Sass, P and Molnar-Perl, I. Simultaneous determination of sugars, sugar alcohols, acids and amino acids in apricots by gas chromatography–mass spectrometry. *Journal of Chromatography A* 847, 91–102 (1999).
15. Fiehn, O, Kopka, J, Dörmann, P, Altmann, T, Trethewey, RN, et al. Metabolite profiling for plant functional genomics. *Nature Biotechnology* 18, 1157–61 (2000).
16. Glassbrook, N, Beecher, C and Ryals, J. Metabolic profiling on the right path. *Nature Biotechnology* 18, 1142–3 (2000).
17. Glassbrook, N and Ryals, J. A systematic approach to biochemical profiling. *Current Opinion in Plant Biology* 4, 186–90 (2001).
18. Trethewey, RN, Krotzky, AJ and Willmitzer, L. Metabolic profiling: a Rosetta Stone for genomics? *Current Opinion in Plant Biology* 2, 83–5 (1999).
19. Hall, RD. Plant metabolomics: from holistic hope, to hype, to hot topic. *New Phytologist* 169, 453–68 (2006).
20. Fritz, C, Mueller, C, Matt, P, Feil, R and Stitt, M. Impact of the C-N status on the amino acid profile in tobacco source leaves. *Plant Cell and Environment* 29, 2055–76 (2006).
21. Kusano, M, Fukushima, A, Redestig, H and Saito, K. Metabolomic approaches toward understanding nitrogen metabolism in plants. *Journal of Experimental Botany* 62, 1439–53 (2011).
22. Abu-Nada, Y, Kushalappa, A, Marshall, W, Al-Mughrabi, K and Murphy, A. Temporal dynamics of pathogenesis-related metabolites and their plausible pathways of induction in potato leaves following inoculation with Phytophthora infestans. *European Journal of Plant Pathology* 118, 375–91 (2007).
23. Allwood, JW, Ellis, DI and Goodacre, R. Metabolomic technologies and their application to the study of plants and plant–host interactions. *Physiologia Plantarum* 132, 117–35 (2008).
24. López-Gresa, MP, Maltese, F, Bellés, JM, Conejero, V, Kim, HK, et al. Metabolic response of tomato leaves upon different plant-pathogen interactions. *Phytochemical Analysis* 21, 89–94 (2010).
25. Mirnezhad, M, Romero-González, RR, Leiss, KA, Choi, YH, Verpoorte, R, et al. Metabolomic analysis of host plant resistance to thrips in wild and cultivated tomatoes. *Phytochemical Analysis* 21, 110–17 (2010).
26. Meyer, RC, Steinfath, M, Lisec, J, Becher, M, Witucka-Wall, H, et al. The metabolic signature related to high plant growth rate in Arabidopsis thaliana. *Proceedings of the National Academy of Sciences of the U S A* 104, 4759–64 (2007).
27. Sreenivasulu, N, Usadel, B, Winter, A, Radchuk, V, Scholz, U, et al. Barley grain maturation and germination: metabolic pathway and regulatory

network commonalities and differences highlighted by new MapMan/ PageMan profiling tools. *Plant Physiology* **146**, 1738–58 (2008).
28. Morcuende, R, Bari, R, Gibon, Y, Zheng, W, Pant, BD, et al. Genome-wide reprogramming of metabolism and regulatory networks of Arabidopsis in response to phosphorus. *Plant Cell and Environment* **30**, 85–112 (2007).
29. Alvarez, S, Marsh, EL, Schroeder, SG and Schachtman, DP. Metabolomic and proteomic changes in the xylem sap of maize under drought. *Plant Cell and Environment* **31**, 325–40 (2008).
30. Charlton, AJ, Donarski, JA, Harrison, M, Jones, SA, Godward, J, et al. Responses of the pea (Pisumsativum L.) leaf metabolome to drought stress assessed by nuclear magnetic resonance spectroscopy. *Metabolomics* **4**, 312–27 (2008).
31. Obata, T, Matthes, A, Koszior, S, Lehmann, M, Araújo WL, et al. Alteration of mitochondrial protein complexes in relation to metabolic regulation under short-term oxidative stress in Arabidopsis seedlings. *Phytochemistry* **72**, 1081–91 (2011).
32. Oliver, MJ, Guo L, Alexander DC, Ryals JA, Wone BW, et al. A sister group contrast using untargeted global metabolomic analysis delineates the biochemical regulation underlying desiccation tolerance in Sporobolus stapfianus. *Plant Cell* **23**, 1231–48 (2011).
33. Sanchez, DH, Siahpoosh, MR, Roessner, U, Udvardi, M and Kopka, J. Plant metabolomics reveals conserved and divergent metabolic responses to salinity. *Physiologia Plantarum* **132**, 209–19 (2008).
34. Shulaev, V, Cortes, D, Miller, G and Mittler, R. Metabolomics for plant stress response. *Physiologia Plantarum* **132**, 199–208 (2008).
35. Trenkamp, S, Eckes, P, Busch, M and Fernie, AR. Temporally resolved GC-MS-based metabolic profiling of herbicide treated plants reveals that changes in polar primary metabolites alone can distinguish herbicides of differing mode of action. *Metabolomics* **5**, 277–91 (2009).
36. Urano, K, Maruyama, K, Ogata, Y, Morishita, Y, Takeda, M, et al. Characterization of the ABA-regulated global responses to dehydration in Arabidopsis by metabolomics. *Plant Journal* **57**, 1065–78 (2009).
37. Riedelsheimer, C, Czedik-Eysenberg, A, Grieder, C, Lisec, J, Technow, F, et al. Genomic and metabolic prediction of complex heterotic traits in hybrid maize. *Nature Genetics* **44**, 217–20 (2012).
38. Riedelsheimer, C, Lisec, J, Czedik-Eysenberg, A, Sulpice, R, Flis, A, et al. Genome-wide association mapping of leaf metabolic profiles for dissecting complex traits in maize. *Proceedings of the National Academy of Sciences* **109**, 8872–7 (2012).
39. Baker, JM, Hawkins, ND, Ward, JL, Lovegrove, A, Napier, JA, et al. A metabolomic study of substantial equivalence of field-grown genetically modified wheat. *Plant Biotechnology Journal* **4**, 381–92 (2006).
40. Chen, YZ, Pang, QY, He, Y, Zhu, N, Branstrom, I, et al. Proteomics and metabolomics of Arabidopsis responses to perturbation of glucosinolate biosynthesis. *Molecular Plant* **5**, 1138–50 (2012).
41. Simoh, S, Linthorst, HJ, Lefeber, AW, Erkelens, C, Kim, HK, et al. Metabolic changes of Brassica rapa transformed with a bacterial

isochorismate synthase gene. *Journal of Plant Physiology* **167**, 1525–32 (2010).
42. Xu, YZ, Santamaria, RdeL, Virdi, KS, Arrieta-Montiel, MP, Razvi, F, et al. The chloroplast triggers developmental reprogramming when MUTS HOMOLOG1 is suppressed in plants. *Plant Physiology* **159**, 710–20 (2012).
43. Fernie, AR and Schauer, N. Metabolomics-assisted breeding: a viable option for crop improvement? *Trends in Genetics* **25**, 39–48 (2009).
44. Kopka, J, Fernie, A, Weckwerth, W, Gibon, Y and Stitt, M. Metabolite profiling in plant biology: platforms and destinations. *Genome Biology* **5**, 109 (2004).
45. Kueger, S, Steinhauser, D, Willmitzer, L and Giavalisco, P. High-resolution plant metabolomics: from mass spectral features to metabolites and from whole-cell analysis to subcellular metabolite distributions. *The Plant Journal* **70**, 39–50 (2012).
46. Maloney, V. Plant metabolomics. *BioTech Journal* **2**, 92–9 (2004).
47. Oksman-Caldentey, K-M and Saito, K. Integrating genomics and metabolomics for engineering plant metabolic pathways. *Current Opinion in Biotechnology* **16**, 174–9 (2005).
48. Stitt, M and Fernie, AR. From measurements of metabolites to metabolomics: an 'on the fly' perspective illustrated by recent studies of carbon–nitrogen interactions. *Current Opinion in Biotechnology* **14**, 136–44 (2003).
49. Sumner, LW, Mendes, P and Dixon, RA. Plant metabolomics: large-scale phytochemistry in the functional genomics era. *Phytochemistry* **62**, 817–36 (2003).
50. Tollenaar, M. and Lee, E. Yield potential, yield stability and stress tolerance in maize. *Field Crops Research* **75**, 161–9 (2002).
51. Sangoi, L. Understanding plant density effects on maize growth and development: an important issue to maximize grain yield. *Ciência Rural* **31**, 159–68 (2001).
52. Takahashi, H. Sulfur metabolism, transport, and signaling. *Annual Review of Plant Biology* **62** (2011).
53. Vance, CP, Uhde-Stone, C and Allan, DL. Phosphorus acquisition and use: critical adaptations by plants for securing a nonrenewable resource. *New Phytologist* **157**, 423–47 (2003).
54. Paponov, I, Sambo, P, Erley, G, Presterl, T, Geiger, H, et al. Grain yield and kernel weight of two maize genotypes differing in nitrogen use efficiency at various levels of nitrogen and carbohydrate availability during flowering and grain filling. *Plant and Soil* **272**, 111–23 (2005).
55. Ainsworth, EA and Bush, DR. Carbohydrate export from the leaf: a highly regulated process and target to enhance photosynthesis and productivity. *Plant Physiology* **155**, 64 (2011).
56. Echarte, L, Rothstein, S and Tollenaar, M. The response of leaf photosynthesis and dry matter accumulation to nitrogen supply in an older and a newer maize hybrid. *Crop Science* **48**, 656–65 (2008).
57. Hudson, D, Guevara, D, Yaish, MW, Hannam, C, Long, N, et al. GNC and CGA1 modulate chlorophyll biosynthesis and Glutamate Synthase (GLU1/Fd-GOGAT) expression in Arabidopsis. *PLoS One* **6**, e26765 (2011).

58. Kant, S, Seneweera, S, Rodin, J, Materne, M, Burch, D, et al. Improving yield potential in crops under elevated CO_2: integrating the photosynthetic and nitrogen utilization efficiencies. *Frontiers in Plant Science* 3, 162 (2012).
59. Miflin, BJ and Habash, DZ. The role of glutamine synthetase and glutamate dehydrogenase in nitrogen assimilation and possibilities for improvement in the nitrogen utilization of crops. *Journal of Experimental Botany* 53, 979 (2002).
60. Amiour, N, Imbaud, S, Clément, G, Agier, N, Zivy, M, et al. The use of metabolomics integrated with transcriptomic and proteomic studies for identifying key steps involved in the control of nitrogen metabolism in crops such as maize. *Journal of Experimental Botany* 63, 5017–33 (2012).
61. Last, RL, Jones, AD and Shachar-Hill, Y. Towards the plant metabolome and beyond. *Nature Reviews Molecular Cell Biology* 8, 167–74 (2007).
62. Trethewey, RN. Metabolite profiling as an aid to metabolic engineering in plants. *Current Opinion in Plant Biology* 7, 196–201 (2004).
63. Kant, S, Bi, YM and Rothstein, SJ. Understanding plant response to nitrogen limitation for the improvement of crop nitrogen use efficiency. *Journal of Experimental Botany* 62, 1499 (2011).
64. Masclaux-Daubresse, C, Daniel-Vedele, F, Dechorgnat, J, Chardon, F, Gaufichon, L, et al. Nitrogen uptake, assimilation and remobilization in plants: challenges for sustainable and productive agriculture. *Annals of Botany* 105, 1141–57 (2010).
65. McAllister, CH, Beatty, PH and Good, AG. Engineering nitrogen use efficient crop plants: the current status. *Plant Biotechnology Journal* (2012) (Epub ahead of print).
66. Oliveira, IC, Brears, T, Knight, TJ, Clark, A and Coruzzi, GM. Overexpression of cytosolic glutamine synthetase: relation to nitrogen, light, and photorespiration. *Plant Physiology* 129, 1170 (2002).
67. Foyer, CH, Lescure, JC, Lefebvre, C, Morot-Gaudry, JF, Vincentz, M, et al. Adaptations of photosynthetic electron transport, carbon assimilation, and carbon partitioning in transgenic Nicotiana plumbaginifolia plants to changes in nitrate reductase activity. *Plant Physiology* 104, 171–8 (1994).
68. Good, AG, Johnson, SJ, De Pauw, MA, Carroll, RT, Savidov, N, et al. Engineering nitrogen use efficiency with alanine aminotransferase. *Canadian Journal of Botany* 85, 252–62 (2007).
69. Shrawat, AK, Carroll, RT, DePauw, M, Taylor, GJ and Good, AG. Genetic engineering of improved nitrogen use efficiency in rice by the tissue-specific expression of alanine aminotransferase. *Plant Biotechnology Journal* 6, 722–32 (2008).
70. Bi, YM, Wang, RL, Zhu, T and Rothstein, SJ. Global transcription profiling reveals differential responses to chronic nitrogen stress and putative nitrogen regulatory components in Arabidopsis. *BMC Genomics* 8, 281 (2007).
71. Lian, X, Wang, S, Zhang, J, Feng, Q, Zhang, L, et al. Expression profiles of 10,422 genes at early stage of low nitrogen stress in rice assayed using a cDNA microarray. *Plant Molecular Biology* 60, 617–31 (2006).
72. Brotman, Y, Riewe, D, Lisec, J, Meyer, RC, Willmitzer, L, et al. Identification of enzymatic and regulatory genes of plant metabolism through QTL analysis in Arabidopsis. *Journal of Plant Physiology* 168, 1387–94 (2011).

73. Gieger, C, Geistlinger L, Altmaier, E, Hrabé de Angelis, M, Kronenberg, F, et al. Genetics meets metabolomics: a genome-wide association study of metabolite profiles in human serum. *PLoS Genetics* **4**, e1000282 (2008).
74. Valdar, W, Solberg, LC, Gauguier, D, Burnett, S, Klenerman, P, et al. Genome-wide genetic association of complex traits in heterogeneous stock mice. *Nature Genetics* **38**, 879–87 (2006).
75. Atwell, S, Huang, YS, Vilhjálmsson BJ, Willems G, Horton M, et al. Genome-wide association study of 107 phenotypes in Arabidopsis thaliana inbred lines. *Nature* **465**, 627–31 (2010).
76. Huang, X, Wei, X, Sang, T, Zhao, Q, Feng, Q, et al. Genome-wide association studies of 14 agronomic traits in rice landraces. *Nature Genetics* **42**, 961–7 (2010).
77. Lu, Y, Zhang, S, Shah, T, Xie, C, Hao, Z, et al. Joint linkage–linkage disequilibrium mapping is a powerful approach to detecting quantitative trait loci underlying drought tolerance in maize. *Proceedings of the National Academy of Sciences* **107**, 19585–90 (2010).
78. Wang, M, Yan, J, Zhao, J, Song, W, Zhang, X, et al. Genome-wide association study (GWAS) of resistance to head smut in maize. *Plant Science* **196**, 125–31 (2012).
79. Suhre, K, Shin, SY, Petersen, AK, Mohney, RP, Meredith, D, et al. Human metabolic individuality in biomedical and pharmaceutical research. *Nature* **477**, 54–60 (2011).
80. Barnes, VM, Teles, R, Trivedi, HM, Devizio, W, Xu, T, et al. Assessment of the effects of dentifrice on periodontal disease biomarkers in gingival crevicular fluid. *Journal of Periodontology* **81**, 1273–9 (2010).
81. Gall, WE, Beebe, K, Lawton, KA, Adam, KP, Mitchell, MW, et al. alpha-hydroxybutyrate is an early biomarker of insulin resistance and glucose intolerance in a nondiabetic population. *PLoS One* **5**, e10883 (2010).
82. Ganti, S, Taylor, SL, Kim, K, Hoppel, CL, Guo, L, et al. Urinary acylcarnitines are altered in human kidney cancer. *International Journal of Cancer* **130**, 2791–800 (2012).
83. Kim, K, Taylor, SL, Ganti, S, Guo, L, Osier, MV, et al. Urine metabolomic analysis identifies potential biomarkers and pathogenic pathways in kidney cancer. *OMICS* **15**, 293–303 (2011).
84. Milburn, MA. Metabolomics for biomarker discovery. *Innovations in Pharmaceutical Technology* **21**, 40–2 (2006).
85. Weiner, J 3rd, Parida, SK, Maertzdorf, J, Black, GF, Repsilber, D, et al. Biomarkers of inflammation, immunosuppression and stress with active disease are revealed by metabolomic profiling of tuberculosis patients. *PLoS One* **7**, e40221 (2012).
86. Yan, J, Jiang, X, West, AA, Perry, CA, Malysheva, OV, et al. Maternal choline intake modulates maternal and fetal biomarkers of choline metabolism in humans. *American Journal of Clinical Nutrition* **95**, 1060–71 (2012).

5

The uniqueness of conifers

Carmen Díaz-Sala, Department of Life Science, University of Alcalá, Spain, José Antonio Cabezas, National Research Institute for Agricultural and Food Technology (INIA), Spain, Brígida Fernández de Simón, National Research Institute for Agricultural and Food Technology (INIA), Spain, Dolores Abarca, Department of Life Science, University of Alcalá, Spain, M. Ángeles Guevara, National Research Institute for Agricultural and Food Technology (INIA), Spain and Mixed Unit of Forest Genomics and Ecophysiology, INIA/UPM, Spain, Marina de Miguel, National Research Institute for Agricultural and Food Technology (INIA), Spain, Estrella Cadahía, National Research Institute for Agricultural and Food Technology (INIA), Spain, Ismael Aranda, National Research Institute for Agricultural and Food Technology (INIA), Spain, and María-Teresa Cervera, National Research Institute for Agricultural and Food Technology (INIA), Spain and Mixed Unit of Forest Genomics and Ecophysiology, INIA/UPM, Spain

DOI: 10.1533/9781908818478.67

Published by Woodhead Publishing Limited, 2013

Abstract: Conifers separated from angiosperms 300 million years ago and currently dominate many of the temperate and boreal forests. This chapter aims to highlight the singularities of this phylum, characterized by very large genomes (18 to 35 Gb) and showing very different structure and composition when compared with angiosperm genomes. Despite an apparently conserved genome structure, conifers demonstrate some competitive capacity, as very different lineages are adapted to a wide variety of environmental conditions. All the above reasons make conifers a unique system for the study of adaptive evolution in plants.

Key words: Conifer genomics, transcriptomics, adaptive capacity, metabolic adaptation.

5.1 Introduction

Conifers belong to the order Coniferales, which, together with Cycadales, Ginkgoales, and Gnetales, constitute the group of gymnosperms. With a fossil record dating from the Carboniferous (c. 310 mya), it was not until the Triassic (245–08 mya) that they radiated. Conifers are considered ancient species. The six conifer families which still exist today appear in the fossil record from this time: Pinaceae (10–11 genera), Podocarpaceae (19 genera), Araucariaceae (three genera), Sciadopityaceae (one genus), Taxaceae (six genera) and Cupressaceae (40 genera), and ca. 653 species (Gernandt et al., 2011).

Gymnosperms diverged from angiosperms in the Carboniferous, about 300 mya or more (Bowe et al., 2000). Flowering plants differentiated enormously, originating a large number of species. On the contrary, gymnosperms split into the four orders right after the divergence from angiosperms, and conifer diversity expanded later during the Mesozoic (ca. 250–65 mya), originating a smaller number of species. For this reason, many biological features of conifers (and gymnosperms) and angiosperms differ greatly, including lifespan, diversification rate, seed morphology, pollination processes, environmental requirements and response to environmental stresses.

Although the total number of species is relatively small, conifers have huge ecological importance. Conifers are the dominant plants in northernmost temperate and boreal ecosystems. Although they are more frequent in the northern hemisphere, they are widespread from

Tierra del Fuego, in South America, to the Arctic Circle, and colonize extreme environments such as high altitudes, ecosystems with nutrient poor soil or with relatively high levels of evaporative demand, and temperate ecosystems that experience seasonal drought. Conifers represent the largest terrestrial carbon sink, with the tallest trees (*Sequoia sempervirens*) reaching 115 m (379 ft), the largest single-stemmed plants by wood volume (*Sequoia dendrongiganteum*) with >1400 m^3, the main stem circumference in diameter (*Taxodium mucronatum*) with almost 58 m (190 ft) at its base, and the longest living non-clonal terrestrial organisms on earth, *Pinus longaeva*, living to ca. 5000 years of age. The typical lifespan of North American conifers has been estimated at 400 years versus the 250 years calculated for angiosperm trees (Loehle, 1988).

Conifers are also of significant economic importance, since they are used as ornamental plants and for timber and paper production. Coniferous forests produce more wood per hectare, and they are preferred for pulp because of the long fibers in the wood. Plantations of conifers for production of wood account for 54 per cent of the area reported in 2005, while broad leaves account for 39 per cent (FAO: State of the World's Forests 2007; *http://www.fao.org/docrep/009/a0773e/a0773e00.htm*). Conifers also provide non-timber products such as resins, which are much appreciated for their chemical properties and their versatile uses, such as the production of varnishes, adhesives, food additives, and cosmetic products, and foodstuffs (particularly seeds). Other important applications are active compounds for the pharmaceutical industry such as taxol, extracted from several species of yews, which disrupts the process of cell division and is used in medical treatment against different kinds of cancer.

Considering the technological progress experienced by most biological disciplines and the enormous amount of information obtained, this chapter highlights important features that support the uniqueness of conifers. These features make them a unique system for unraveling pathways and processes controlling traits such as longevity and height, as well as the broad ecological ranges covered by many of these species, which imply sophisticated adaptive responses.

5.2 Functional differentiation

It has often been postulated that a more advanced reproductive system in angiosperms is one of the main reasons for their ecological success over gymnosperms, which carry naked seeds, since the Cretaceous (Crepet and Niklas, 2009). Conifers have nevertheless retained some competitive capacity in the last 100 million years due to their particular adaptive response to climate change (see Brodribb et al., 2012, for a comprehensive review).

Indeed, the long evolutionary history of conifers has seen the development of very different lineages adapted to a wide variety of environmental conditions. From tropical to almost semi-arid deserts, species from families such as Pinaceae or Cupressaceae are found covering an ample range of climate conditions. This translates into a rich plethora of functional strategies, although it should be pointed out that there is a suite of common traits as a consequence of the monophyletic origin of the group (Brodribb et al., 2005; Figure 5.1).

Specific xylem anatomies (Sperry and Sullivan, 1992) and needle or scale-leaves with their particular leaf hydraulic function (Brodribb et al., 2010; Zwieniecki et al., 2006) are two specific morphological and functional traits that define the particular lifestyle of conifers. These particular features are present along with others related to their long lifespan or the height or volume reached by some species. Probably these traits, as well as the broad ecological ranges covered by many species, have conditioned many of the genetic, functional and ecological characteristics of the species that constitute the group (Brodribb et al., 2005; Maherali et al., 2004).

The very different species seem to have maintained, through their long-standing evolution, a very conservative wood anatomy system for water

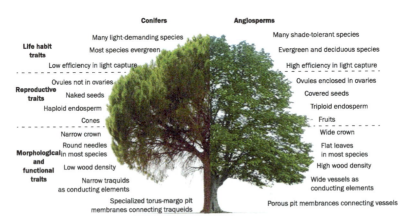

Figure 5.1 Key biological differences between conifers and angiosperms

transport based on narrow conduits (tracheids). The high resistance to stresses, such as drought or low temperature, is explained by the development of a xylem especially resistant to the loss of functionality from the cavitation phenomena associated with these stresses (Sperry and Sullivan, 1992; Sperry et al., 2006; Willson and Jackson, 2006). The tracheids of many conifer species have a small diameter, which, together with the permeability of pit membranes that interconnect conducting elements, makes the xylem of conifers highly resistant to tensions generated as a consequence of water stress or freeze-thaw cycles (Sperry and Sullivan, 1992). In fact, conifer tracheids maintain a complicated equilibrium between hydraulic sufficiency and resistance to embolisms formed as a consequence of the tension established within the xylem (Pittermann et al., 2006). Recent reports seem to point to specific features of the elements included in the pits as the main control points in the sensitivity of the hydraulic system to cavitation (Cochard et al., 2009; Pittermann et al., 2010). In this respect, the presence of a torus-margo pit membrane system in the more evolved conifer tracheids is an important evolutionary achievement of many conifers, making them as competitive as current angiosperms from a hydraulic point of view (Becker et al., 1999; Pittermann et al., 2005).

Conifers possess leaves with a single-vein hydraulic system in most genera, with typical needle or imbricate scale leaf forms (Brodribb et al., 2010). From an ecological point of view, many species of the Pinaceae (*Pinus* sp. and *Cedrus* sp.) and Cupressaceae (e.g. *Juniperus* sp. and *Cupressus* sp.) are high-light demanding for growth. This heliophilous habit would require leaf hydraulic systems with a high efficiency to sustain the high transpiration fluxes common in sunny environments, but on the other hand constraining the size and form of leaves (Zwieniecki et al., 2004).

Despite conifers being an old lineage and comprising a lower number of species than angiosperms, remarkably high intra-specific variability is observed in some functional traits of some species of genera, such as *Pinus* sp. (Aranda et al., 2010; Sánchez-Gómez et al., 2011), *Pseudotsuga* sp. (Zhang et al., 1993), *Picea* sp. (Flanagan and Johnsen, 1995) and *Larix* sp. (Zhang and Marshall, 1994). In this respect significant differences have been reported not only at population level (Ducrey et al., 2008; Nguyen-Queyrens and Bouchet-Lannat, 2003) but also at family level (Johnsen et al., 1999) and even between clones of a full-sib family (de Miguel et al., 2012) in traits related to growth, gas exchange, water use efficiency, hydraulic system or osmotic regulation. This intra-specific genetic richness is mirrored by a high variability at the molecular level, from genes to chemical composition.

5.3 Genome structure and composition

Conifers have relatively large genomes when compared with angiosperms (Figure 5.2). Their genome sizes range between the 6.5 Gb of *Lepidothamnus intermedius* and the 37 Gb of *Pinus ayacahuite* (Plant DNA C-values Database; *http://data.kew.org/cvalues/CvalServlet?querytype=1*; Murray et al., 2010).

Although several angiosperms may have larger genomes, the mode of their genome sizes is significantly lower than in conifers (<1Gb vs. >15 Gb; Ahuja and Neale, 2005). Analyses of conifer diversification and genome size evolution show a dynamic evolutionary process contrasting with their reputation as ancient and relict species. There is also strong evidence for a recent and large increase in the rate of genome size evolution in *Pinus*, unique among gymnosperms (Burleigh et al., 2012). When compared with angiosperms, pairwise comparisons of spruce and pine orthologs revealed a considerably slower rate of nucleotide substitution in conifers, on average 15-fold, but a significantly higher dN/dS (threefold higher non-synonymous distance/synonymous distance) ratio (Buschiazzo et al., 2012). This result may indicate higher adaptation and clearly points to the uniqueness of conifer genome evolution. In any case, the patterns of the evolution of conifer genomes and the question of

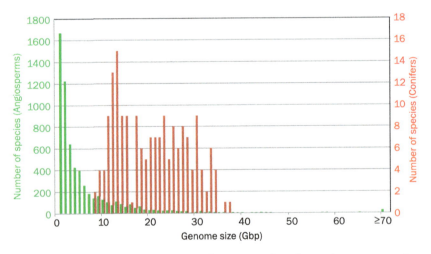

Figure 5.2 The distribution of genome sizes in angiosperms and conifers. Angiosperm data in green (left y-axis) and conifer data in red (right y-axis)

why they have such large genome sizes remain unsolved, although sequencing projects currently running are giving us new insights.

The number of chromosomes is much less variable in conifers (2n=18–66) than in angiosperms (2n=4–640), as well as their size (5-15μm and >0.5 to <30μm, respectively; Leitch and Leitch, 2012). For example, pine chromosomes are uniform among all species in both number (2n=24) and appearance, although individuals with accessory chromosomes can be found in many of them (Ahuja and Neale, 2005; Sedel'nikova et al., 2011). Much chromosome number diversity in angiosperms is associated with polyploidy (whole genome duplications). In fact, polyploidy (the presence of more than two genomes per nucleus) is one of the main sources of variation in genome size in angiosperms. It has been estimated that 50–80% of angiosperms are polyploids (Masterson, 1994; Otto and Whitton, 2000). Polyploidy does not play a similar role in conifers, since only three cases are known, all of them in the Cupressaceae family (Ahuja and Neale, 2005).

Genome complexity can be estimated by methylation–filtration and CoT analysis, genome-filtration techniques for constructing reduced-representational libraries for genome sequencing (Springer et al., 2004). These techniques have revealed that a large fraction of the conifer genome is composed of repetitive sequences (>75%; Kovach et al., 2010; Morse et al., 2009).

The sequencing of large-insert 'bacterial artificial chromosome' (BAC) genomic libraries for several conifer species allows us to go more deeply into the study of conifer genomes. The majority of genes are single or low copy in nature, and conifers appear to have functional gene numbers similar to those of diploid angiosperms (Rigault et al., 2011; Ritland, 2012). However, the low-repetitive and single-copy component greatly exceeds the amount of genomic information that is sufficient for full biological functionality in known small-genome angiosperms (Morgante and De Paoli, 2011). For example, the 7.4% estimated for bald cypress (*Taxodium distichum* var. *distichum*) is 4.6 times the size of the Arabidopsis genome (Liu et al., 2011). The large amount of single and low-copy DNA is likely the result of divergence among ancient repeat copies and gene/pseudogene duplication. In this way, Kovach et al. (2010), examining ten *Pinus taeda* BACs, found that pseudogenes were five times more common than genes with potential protein coding functions. And Garcia-Gil (2008) found that in Scots pine (*Pinus sylvestris*) pseudogenes were much more numerous than intact coding genes in the phytochrome gene family. These studies point towards most conifer genes having many paralogues or pseudogenes, as well as genes that appear to be conforming gene islands, quite compact relative to the dauntingly vast genome size.

A more detailed analysis of the repetitive fraction of the conifer genome revealed that under stringent conditions of hybridization only 24% of *Pinus taeda* genome is nearly identical, but under more permissive conditions up to 80% of its genome is repetitive (Kovach et al., 2010). This result suggests the existence of a massive 'low-copy' fraction and a very small 'high-copy' fraction, containing a few repeat families and occurring fewer than 100 000 times in the 22-Gb genome. Morgante and De Paoli (2011) estimated that retroelements and non-characterized repetitive elements account for more than 75% of the *Picea abies* genome. Even larger values (90.5%) have been estimated for *Taxodium distichum* (Liu et al., 2011). These large proportions of repetitive DNA point towards an evolution of genome sizes related to retrotransposons that are limited to conifers (Burleigh et al., 2012; Elsik and Williams, 2000; Magbanua et al., 2011; Morse et al., 2009). Retrotransposons are widespread across the chromosomes of gymnosperms; 42% of the *Picea abies* genome consists of long terminal repeat (LTR) retrotransposons (Morgante and De Paoli, 2011). Different types of LTR retrotransposons have been detected in conifer genomes, including Ty1-copia and Ty3-gypsy. Recently, Gymny, a new family related to Arabidopsis Athila elements, has been detected only in *Pinus*. Gymny has been introduced or amplified more recently than other retrotranspons in gymnosperms. The study of ten *P. taeda* BACs suggests that isolated repetitive elements such as LTR retrotransposons can be discerned from a background of fragmented fossil repeats of unknown origin (Kovach et al., 2010). In this way, conifer genomes can be distinguished from angiosperm genomes by the old age and high degree of divergence in both their intact and fragmented LTR retrotransposons, compared with the younger LTR retroelements in angiosperm genomes (Morgante and De Paoli, 2011). Detailed analysis of conifer BAC sequences has allowed the estimation of retroelement insertion dating back to the differentiation of Pinaceae species, long before the main episodes of genome expansion in angiosperms. As pointed out by Morgante and De Paoli (2011), this result suggests different evolutionary consequences for genome diversity and flexibility in perennial conifers. Kovach et al. (2010) found only two conifer-specific LTR retroelements that may still be active after 140 million years. In addition, the IFG7 family of gypsy-like elements appears to be active in both subgenera of *Pinus*, which diverged approximately 110 million years ago. Moreover, Magbanua et al. (2011) and Morse et al. (2009) found that IFG-7 and Gymny alone account for about 5.8% and 0.7% of total nuclear DNA in *Pinus taeda*, respectively (1.26 Gb and 157 Mb, equivalent to about eight and one *Arabidopsis thaliana* genomes, respectively).

Comparative genetic mapping highlights similarities and differences in genome content and genome organization between species. Genetic maps constructed in one species can be compared with those developed in closely related species by using common markers. Map comparisons are based on the study of orthologous markers, such as Conserved Ortholog Sets (COS) markers, which are single-copy evolutionarily conserved genes in two or more species that share common ancestry (Fulton et al., 2002). Liewlaksaneeyanawin et al. (2009) identified a set of 239 COS markers useful for synteny analyses between *Picea, Pinus* and *Pseudotsuga* species. But they found that these markers have low genetic variation, often being completely homozygous for the polymerase chain reaction (PCR) product of a mapping parent, precluding their use in mapping for certain pedigrees. In any case, comparative mapping in the Pinaceae provides evidence of a high level of macro-synteny and macro-co-linearity (marker content and order between linkage groups, respectively) in conifers (Brown et al., 2001; Chagné et al., 2003; Devey et al., 1999; Jermstad et al., 2010; Komulainen et al., 2003; Krutovsky et al., 2004; Pelgas et al., 2005; Shepherd and Williams, 2008). Nevertheless, several exceptions have been found, such as a chromosomal fission and an inter-chromosomal rearrangement in the genome of *Pseudotsuga menziesii* and a breakdown in synteny in *Picea* (Krutovsky et al., 2004; Pelgas et al., 2005, 2006). In addition, the success in the alignment of maps suggests that the Pinaceae genome macro-structure closely resembles the macro-structure it had more than 100 million years ago, at the time of the split between *Pinus* and the other major lineages of the Pinaceae. This is a remarkable feature in the plant kingdom, and suggests that the pine family can be viewed as one genetic system, allowing genomic information to be readily transferred, in contrast to angiosperm species with even one-tenth evolutionary separation (Ritland et al., 2011).

The chloroplast is a key organelle, whose genome contains genes involved in photosynthesis. Chloroplast DNA structure and evolution also clearly seem to differ between conifers and angiosperms. Conifer chloroplasts show a different inheritance when compared with angiosperms, since they are paternally inherited in the majority of conifers studied, while among the angiosperms inheritance is predominantly maternal (Neale and Sederoff, 1989). However, little is known about the biological implications of this genetic strategy, which may favor an increased gene flow and therefore variability from distant pollen donors, or about nuclear genes controlling plastid inheritance. Full chloroplast genome sequences are available at the GeneBank for 12 conifer species. Conifer chloroplast DNA (cpDNA) is smaller (approximately 120–35 kb) than in angiosperms (140–220 kb), due to the lack of several elements that characterize most land plants (Lin et al., 2010). Structural comparisons of cpDNAs revealed differential loss of two large inversions (21 and 42 kb; Strauss et al., 1988; Wu et al., 2011) in two groups of conifers. The structural analysis of these inversions revealed that Pinaceae and Cupressophytes retain different residual interspersed repeat (IR) copies. Also, most of the conifer cpDNAs have also lost all 11 functional genes (ndh genes) for subunits of a putative NADH dehydrogenase, maybe due to the ancestral plastid NDH genes' transfer into the nucleus (Wakasugi et al., 1994). However, this deletion does not occur in all conifer species; NDH is present in the chloroplast genomes of *Cryptomeria* and *Cycas* (Hirao et al., 2008; Wu et al., 2007).

5.4 Genome function

5.4.1 Transcriptome characterization

Understanding the biological principles guiding conifer biology requires, among other things, the knowledge of transcriptional responses occurring in multiple cell types, tissues, organs and developmental stages of an individual. Several recent studies are moving us towards this understanding. Transcriptomics attempts to measure the transcript levels of all genes from a given genome. In the post-genomics era, comprehensive characterization of conifer transcriptomes will increase our understanding of the complex regulatory networks associated with the phenotypic variation of many traits, and will allow us to decipher the connection between genomes and phenotypes. High-throughput technologies that do not depend on prior probe selection, such as next generation sequencing-based technologies, have revolutionized functional studies by offering the potential to interrogate transcriptomes of non-model organisms in detail. These technologies provide new opportunities, not only to analyze the dynamics of gene expression underlying conifer responses to developmental or environmental changes, but also to analyze non-coding RNAs and to clarify aspects of epigenetic regulation of gene expression.

Large-scale cDNA sequencing projects from conifers have been performed for over a decade as a first approach to the study of conifer transcriptomes. Sanger sequencing of EST collections from *Pinus* sp., *Picea* sp., *Pseudotsuga* and *Cryptomeria* cDNA libraries resulted in over a million EST sequences publicly available using these methodologies (Futamura et al., 2008; Kirst et al., 2003; Lorenz et al., 2006; Liang et al., 2007; Pavy et al., 2005; http://www.ncbi.nlm.nih.gov/nucest/). New techniques for cDNA synthesis resulting in full-length cDNA-enriched libraries as well as clustering of ESTs obtained by next generation sequencing technologies have improved the characterization of the coding regions to predict gene function and the coverage of the gene catalog (Futamura et al., 2008; Lorenz et al., 2012; Parchman et al., 2010; Prunier et al., 2011; Ralph et al., 2008; Rigault et al., 2011; Ritland, 2012; http://dendrome.ucdavis.edu/treegenes/; http://www.ncbi.nlm.nih.gov/sra).

A reference transcriptome for *Picea glauca* has recently been reported (Rigault et al., 2011). Current projects will supply more transcriptomes from other conifers in a short period of time, which will allow comparative analyses (http://www.scbi.uma.es/sustainpine/; http://pinegenome.org/pinerefseq/; http://www.congenie.org/; http://www.procogen.eu).

The availability of a *Picea glauca* reference transcriptome has already made it possible to conduct the first comparative analyses with angiosperm transcriptomes to begin to describe the extent of conservation and divergence between a conifer and angiosperms. Estimations of gene content indicate that the transcriptome size is at the low end of the angiosperm range (Rigault et al., 2011). A comparison of the encoded proteins showed that many known protein domains that have been described in angiosperms are conserved in conifers (Bedon et al., 2010; Côté et al., 2010; Rigault et al., 2011; Sánchez et al., 2007; Solé et al., 2008; Vielba et al., 2011), and phylogenetic analyses of gene families showed specific evolutionary patterns (Mackay and Dean, 2011). An example of these different evolutionary paths is the evolution of enzymes involved in diterpenoid metabolism (Chen et al., 2011; Hamberger et al., 2011; Keeling et al., 2010, 2011). The conserved general diterpenoid metabolism of ent-kaurenoic acid biosynthesis, with no obvious gene duplication in conifers, is indicative of tight control of general plant hormone metabolism. Specialized metabolism, in contrast to general metabolism, is permissive of redundancy and diversification, which may serve as an adaptive advantage, for example, in the chemical defense of sessile, long-lived conifer trees (Chen et al., 2011). The situation in conifers, in which both bifunctional and monofunctional diTPS (diterpene synthase) are found together but with separate roles in specialized and general metabolism, is different from any angiosperm species that has been characterized for parallel general and specialized diterpenoid biosynthesis. There is no report of both monofunctional and bifunctional diTPS in any angiosperm system (Hamberger et al., 2011; Keeling et al., 2010, 2011). Aminocyclopropane carboxylic acid synthase (Ralph et al., 2007a), dirigent proteins (DIR) and *DIR-like* genes (Ralph et al., 2007b), asparagine synthetase (Canales et al., 2012), stilbene synthase and chalcone synthase (Hammerbacher et al., 2011), specific genes within the phenylpropanoid pathway (Porth et al., 2011), and the HD-Zip III (Coté et al., 2010; Rigault et al., 2011) and R2R3-MYBs (Bedon et al., 2010) families of transcription factors also highlight distinct evolutionary trajectories of angiosperm and gymnosperm plants. These different evolutionary paths could have led to the loss of certain functions as well as neo-functionalization or sub-functionalization within the angiosperms.

5.4.2 Expression analyses

The development of microarray technology has enabled the simultaneous examination of thousands of genes, thus providing a comprehensive view of gene activity. Microarray gene expression data now cover several developmental events, as well as responses to a variety of environmental changes in conifers. Genes involved in xylem development and wood formation, biotic and abiotic stresses as well as other developmental processes have been extensively studied in conifers, mainly using microarray technology.

The xylem transcriptome is highly conserved in conifer species, but the nucleotide sequences of the xylem transcriptome are significantly distinct among angiosperms; thus considerable evolution of the xylem transcriptome has occurred in angiosperms (Li et al., 2010; Paiva et al., 2008). The conifer-specific xylem unigenes may share the functions of related homologs, but some may have unique roles in gymnosperm wood formation compared with angiosperms (Li et al., 2010).

Another set of studies has examined the transcriptome and the dynamics of expression of responses to insect attack. Genes involved in the synthesis of terpenes and phenolic compounds seem to play an important role in conifer defense (Hamberger et al., 2011; Porth et al., 2011; Ralph et al., 2007a); however, Verne et al. (2011) found few of these genes to be differentially expressed between putatively resistant and putatively susceptible spruce trees. These authors identified that putative small heat-shock proteins (sHSP) and several other stress-related proteins seem to be down-regulated in resistant trees.

Drought and cold stresses represent major abiotic stresses faced by conifers. Gene expression was evaluated in *Pinus taeda* roots following drought stress, drought recovery and watered roots by Lorenz et al. (2011). Genes commonly associated with drought response in pine and other plant species (Nilson and Assmann, 2010), as well as a number of abiotic and biotic stress-related genes, were up-regulated in drought-stressed roots. Perdiguero et al. (2012) also identified, by suppression subtractive hybridization and microarray analysis, drought-stress candidate genes belonging to different functional groups. Results indicate similarities between mechanisms of drought tolerance in the roots of gymnosperms and angiosperms. Similarly, Joosen et al. (2006) identified genes associated with cold tolerance that were associated with this response in other species.

Gene expression profiling of other developmental processes such as developmental transitions in apical shoots (Friedmann et al., 2007), including seasonal variations (Yang and Loopstra, 2005) and seasonal bud burst (Asante et al., 2009), nitrogen metabolism (Canales et al., 2010), embryo development (Cairney et al., 2006), adventitious organogenesis (Abarca and Díaz-Sala, 2009a, b; Alonso et al., 2007; Brinker et al., 2004; Sánchez et al., 2007; Solé et al., 2008), and synthesis of natural products (Lenka et al., 2012) have also been studied. Genes showing similarities to their angiosperm homologues have been described (Nilsson et al., 2007).

Despite the power of microarrays, they are limited in providing relative abundance information about identified genes and gene models. Therefore, deep sequencing of transcripts (RNA-seq) provides an alternative to microarray technology. Additionally, because RNA-seq does not depend on genome annotation, RNA-seq has emerged as the method of choice for transcriptional profiling in non-model organisms. As a new technology, there are unique challenges that come with analyzing RNA-seq data, including developing methods, algorithms and pipelines (e.g. library preparation procedures, RNA quantification, isoform detection and quantification, etc.). Despite these challenges, and because of the improved throughput and lower cost, RNA-seq is already shedding light on the complexity and regulation of the conifer transcriptome (Mackay and Dean, 2011).

Cell-type specific profiling reveals more transcriptional complexity than whole organ profiling, because of dilution of cell-type specific genes in whole organ experiments. Methods are described that allow analysis of RNA, enzyme activity and metabolites in individual tissues isolated by laser microdissection from woody conifer stems (Abbott et al., 2010;

Hamberger et al., 2011). Combined analysis of transcripts, proteins and metabolites of individual tissues will facilitate future characterization of complex processes of woody plant development, including periodic stem growth and dormancy, cell specialization and defense, and may be applied widely to other plant species and developmental processes.

5.4.3 Small RNA

Together with regular gene transcripts, mRNA isolation yields a population of small-sized non-coding RNAs that contribute to transcriptional and post-transcriptional regulation of gene expression and appear to be involved in plant development and responses to stress. Two types of small RNAs (sRNAs) have been characterized in plants: short-interfering RNAs (siRNAs) and micro RNAs (miRNAs). siRNAs are 21–22 nucleotide (nt) molecules that are synthesized from double-stranded RNA, either from viral RNAs or from endogenous antisense RNA molecules. miRNAs are RNA molecules processed from a specific group of transcripts that contain a self-complementary sequence and form a stem-loop structure. In angiosperms, two populations of miRNAs have been identified: 21 nt miRNAs, many of which are well conserved across different taxonomic groups, and 24 nt miRNAs, which are more abundant and have a higher sequence diversity and a higher level of taxonomic divergence (Sun et al., 2011).

In most eukaryotes, sRNAs are processed by DICER-like RNAses (DCL) and incorporated into an RNA-induced silencing complex (RISC; Bernstein et al., 2001). RISC complexes containing 24 nt miRNAs induce silencing by histone and DNA modifications in repetitive regions, so they are involved in heterochromatin formation to control transposable elements (TE). In general, siRNAs and 21 nt miRNAs control gene expression either by inducing mRNA cleavage or translational inhibition of specific target transcripts that are recognized by sequence homology or by inducing changes at the chromatin level (Dolgosheina et al., 2008; Kidner and Martienssen, 2005).

Pre- and post-transcriptional regulation of gene expression by sRNAs affects most developmental processes, and appears to be important for basic processes such as induction of polarity, radial patterning, meristem function and organ boundary formation (Kidner and Martienssen, 2005; Lu et al., 2008). It is also important for developmental switches such as phase change and floral induction, and for environmentally induced changes such as circadian or seasonal cycles. In general, these processes are regulated by highly conserved 21 nt miRNAs that target conserved transcription factors. Populations of 21 nt miRNAs with higher sequence diversity have been associated with stress responses and have homology with stress-related genes (Lu et al., 2008).

Most of the plant sRNA sequences that are known have been identified in angiosperms, either by computational prediction of miRNA-encoding genes in the genomic sequence or by direct sequencing of sRNA populations. In contrast, information on sRNAs from gymnosperms is sparse, because no genomic sequence is available yet and only a small number of studies have been published so far. However, the available data show that conifer sRNA populations have some particularities that might indicate an evolutionary divergence in ancient, highly conserved processes.

SiRNAs appear not to be present in sRNA populations from conifers. A detailed study of the size distribution of miRNAs showed that the population of 24 nt miRNAs is very small, in contrast with angiosperms, in which 24 nt miRNAs are the most abundant sRNA species. In conifers, the main miRNA population is the 21 nt miRNA, and only a small fraction of the molecules represent conserved members of miRNA families described in other plant groups. The great majority of the 21 nt miRNA molecules have a high degree of variation at the sequence level, which contrasts with the situation in angiosperms, in which 21 nt miRNAs are highly conserved (Dolgosheina et al., 2008; Sun et al., 2011).

The highly variable population of 21 nt miRNAs in conifers suggests a more dynamic role in adaptation to stress. In fact, specific miRNAs have been related to transgenerational stress responses via signaling from maternal tissues to the developing embryo, a process that involves epigenetic changes at the chromatin level and provides a source for adaptive plasticity that could be essential in conifer species (Yakovlev et al., 2010).

The absence of 24 nt miRNAs suggests a specific mechanism for the control of TE. DCL3, the specific DCL isoform that processes 24 nt in angiosperms, has not been found in conifer transcript collections so far. In addition, a new DCL family that appears to be specific to conifers has been identified, suggesting that conifers could have developed a specific TE silencing mechanism, using 21 nt miRNAs and a specific DCL isoform (Dolgosheina et al., 2008).

It is important to take into account that the proportion of transcribed TE that is detected in cDNA samples from conifers is usually higher than in other plant groups (Parchman et al., 2010). In addition, the dynamics of the evolution of the TE population at the genomic level appear to be unusual, with a steady and slow accumulation of TE sequences, no removal of older elements and a strong degree of sequence conservation across gymnosperms, from Ginkgoaceae to Pinaceae. This, together with the differences related to the miRNA size, suggests a potential mechanism for the evolution of their unusually large genomes, which include, as mentioned previously, over 42% of repetitive DNA consisting of LTR in heterochromatin (Morgante and De Paoli, 2011).

5.5 Chemical divergence

Plants produce an amazing variety of low molecular weight compounds, probably hundreds of thousands. Only a few of them are part of the primary metabolic pathways (those common to all organisms). The rest are called secondary metabolites, defined as compounds whose biosynthesis is restricted to selected groups of plants. When angiosperms and gymnosperms split, new metabolic pathways had to be developed from existing primary metabolism. Since then, the gymnosperm has conducted its particular metabolic adaptation to address the problems of survival and reproduction, constantly exploring new tricks needed to get enough nutrients, manage water availability and light intensities, care for its reproduction and protect against pests, often providing different chemical solutions to common problems in different plant lineages.

One component of the chemical defense system of both angiosperm and gymnosperm is the production of exudates such as resins, gums, gum resins, or kinos, and in conifers these exudates are terpenoid oleoresin and terpenoid volatile emissions (Keeling and Bohlmann, 2006a, b; Phillips and Croteau, 1999; Zulak and Bohlmann, 2010). For this, conifers have specialized anatomical structures for the accumulation of resin terpenes, which can be as simple as the resin blisters found in *Abies* spp., or more complex, such as the resin-filled canals of *Picea* spp. and *Pinus* spp. that are interconnected in a three-dimensional reticulate system (Martin et al., 2002).

Resins and volatile-rich resins are mainly composed of terpenoids and restricted to conifers (Pinophyta), whereas gums are composed of polysaccharides and are produced by some conifer families (including Araucariaceae and Podocarpaceae), non-conifer gymnosperms, and angiosperms (Lambert et al., 2010; Tappert et al., 2011). This implies that the production of terpenoid-based resins is a defining metabolic characteristic of the conifers. On the basis of their micro-Fourier transform infrared spectra, Tappert et al. (2011) subdivided conifer resins into two distinct resin types that reflect compositional differences in their terpenoid constituents: pinaceous and cupressaceous resins. Pinaceous resin is produced by members of the Pinaceae and consists mainly of diterpenes that are based on abietane/pimarane skeletal structures. Cupressaceous resin is associated with members of the Cupressaceae, Sciadopityaceae, Araucariaceae, and Podocarpaceae and consists mainly of diterpenes that are based on the labdanoid structures. Volatile-rich resins are found exclusively within the Pinaceae, reflecting a generally higher abundance of mono- and sesquiterpenoid volatiles in resins of this family.

Terpenoids in conifers are not restricted to the exudates, but can be found in needles, knotwood, bark, wood, branches and roots, showing qualitative differences between species (such as diterpene-taxol and other taxanes, only detected in the different tissues from *Taxus*) (Elsohly et al., 1995), within species (such as two different terpenoid profiles in needles from *Pinus pinaster*) (Arrabal et al., 2012), between tissues (such as the absence of dehydroabietadienol and sandaracopimaredienol in xylem and wood of *Picea sitchensis*, and its abundance in bark and phloem) (Hamberger et al., 2011), and from year to year, as well as related to environmental circumstances (such as the absence of sabinene in needles from *Juniperus communis* at low latitudes, and its significant increase at higher latitudes) (Martz et al., 2009). Besides qualitative differences, numerous quantitative differences in the terpenoid profiles from conifers related to these same factors have been described.

These terpenoids are produced enzymatically by terpene synthases (TPSs), and a gymnosperm-specific TPS subfamily is recognized (Chen et al., 2011), based on the phylogeny and function of TPSs, including mono-, sesqui- and diterpene synthases (Martin et al., 2004), as well as the single-product and multi-product enzymes of conifer diterpene resin acid biosynthesis (Keeling et al., 2008; Peters et al., 2000) and taxadiene synthase from *Taxus* (Wildung and Croteau, 1996). In addition, conifer-specific cytochrome P450 monooxygenases (CYP720B family) produce oxidation at the C-18 carbon of ring A in diterpenes, induced by insect attack on trees, leading to the corresponding diterpene alcohols, aldehydes, and acids (Hamberger et al., 2011; Nelson and Werck-Reichart, 2011; Zulak and Bohlmann, 2010). Other conifer-specific CYP families are CYP725, CYP750, CYP798 and CYP799. No function has been assigned yet to the last two. CYP725 is in the taxol biosynthesis pathway (Jennewein et al., 2004; Rontein et al., 2008): CYP725A is only found in the *Taxus* genus, while CYP725B is also found in *Pinus* and *Picea*.

Flavonoids, phenols, stilbenes, waxes, fats, tannins, and sugars are also among the many classes of compounds known as secondary metabolites that have been detected in conifers. The range of compounds is extensive, with wide-ranging chemical, physical, and biological activities. The concentration of these metabolites in conifers is also affected by species, type of tissue, age, seasonal time, environmental growing conditions, and so on, also showing qualitative and quantitative differences from non-conifers (Stevanovic et al., 2009). Thus, lignin in coniferous gymnosperms does not contain syringyl units, which makes it different from lignin of many other vascular plants, including angiosperms (for a review, see Harris, 2005), causing the classical division into softwoods and hardwoods according to their different physical–mechanical properties. In general, softwoods are richer than hardwoods in lignans, closely related to lignins. Particularly, conifer knotwoods are a very rich source of lignans, maybe even the richest among all trees, and one of the richest in all nature (Holmbom et al., 2003; Smeds et al., 2007), with levels 50% higher than in a normal wood. Their levels in *Abies, Pseudotsuga, Larix*, and especially in *Picea*, are highlighted, with 7-hydroxymatairesinol as the major lignan (over 65–85%) in *Picea abies* knot extracts, processed as a form of food supplement in Finland. Lariciresinol and secolariciresinol have been identified as major lignans in knotwood from *Abies and Pinus*, although lower levels have been detected in pines (Willför et al., 2003).

Other compounds, such as stilbenes, occur in many vascular plants, but are principally found in conifer bark. Among them, resveratrol glucoside (piceid) has been identified as a *Picea abies* bark constituent, along with the stilbene-aglycones resveratrol, astringin and isohapontin, while pinosylvinmonomethyl ether and other pinosylvin derivatives have been found in extracts from various pines: *Pinus strobus, P. banksiana, P. contorta, P. cembra, P. sibirica*, and *P. sylvestris* (Pietarinen et al., 2006). A particularly high amount of stilbenes has been determined in knotwood of *Pinus sibirica*, about 46% of total hydrophilic extract (Willför et al., 2003). Among knotwood extracts from different conifers (*Abies, Pseudotsuga, Thuja, Picea, Larix*), only members of *Pinus* were found to contain stilbenes, which are also regular constituents of pine heartwood (Willför et al., 2003).

The tannins detected in conifers are condensed tannins, since hydrolysable ones are limited to dicotyledonous plants. They are very abundant in the bark. The most studied is *Pinus* spp. bark, since a proanthocyanidin-rich extract obtained from the bark of *P. pinaster* growing in the Landes de Gascogne region (France) is marketed under the names of Pycnogenol^R and Oligopin^R, as a food supplement and as a remedy for cardiovascular diseases. These extracts are based primarily on complex mixtures of oligomeric and polymeric procyanidins (from two to seven units), containing also the monomeric flavanols (-)-catechin, (-)-epicatechin, and (-)-catechingallate. Species and geographical location of growth cause great variability of the amounts of active compounds in the extracts (Packer et al., 1999; Yesil-Celiktas et al., 2009). These compounds are also abundant in *Abies, Taxus*, and *Cedrus* bark extracts (Willför et al., 2009).

Alongside condensed tannins, other flavonoids have been found in conifers, especially in needles (Cannac et al., 2011; Kaundun et al., 2000; Slimestad, 2003). These include flavonols, such as myricetin, quercetin, larycitrin, kaempferol, isorhamnetin, and syringetin, and also flavones and dihydroflavonols, together with dihydrochalcones and flavanones. At the macrosystematic level (from genus to order), the abundance of proanthocyanidins (including prodelphinidin) and the absence of C-glycoflavones allow distinction of pines from the other Pinales; they are themselves characterized by the absence of biflavones, present in Cupressaceae and Araucariaceae (Lebreton, 1990). In addition, qualitative and quantitative differences were found related to genus, species, and clones. Thus, as an example, flavonol acetyl-glucosides were only found in *Picea* and *Abies*, while flavonol-rhamnosides were present only in *Abies* spp. Qualitative and quantitative differences related to environmental conditions are also important, since flavonoids are highly indicative of plant responses to environmental biotic and abiotic factors.

5.6 Meeting the challenge: the system biology approach to unraveling the conifer genome

The development and fast evolution of next-generation sequencing (NGS) technologies have made it possible to address sequencing of enormous genomes such as those of conifer species. The combination of these techniques, which progressively provide an increasing number of longer reads, with strategies that reduce the complexity of conifer genomes, such as the use of haploid DNA extracted from megagametophytes (maternal tissue surrounding the embryo), is a strategy followed by different ongoing conifer genome sequencing projects. Nowadays, the use of different size insert mate-pair (2–20 kb) and paired-end (180–800 bp) libraries, combined with fosmid pool-based and fosmid-end NGS sequencing, may feasibly achieve 50× coverage based on paired-end sequencing, resulting in reads that will be assembled in unordered contigs, as well as 50× coverage based on shotgun fosmid sequencing, which will improve scaffold size, providing significant improvement over the draft assembly.

Genome sequencing of at least eight conifer species (*Picea abies, Picea glauca, Pinus taeda, Pinus lambertiana, Pinus elliottii, Pinus pinaster, Pinus sylvestris* and *Pseudotsuga menziesii*) is approached in the frame of different European, US and Canadian projects (Mackay et al., 2012). Close collaboration among groups working on conifer genomics will ensure a rapid advance to achieve genome drafts, in some cases applying re-sequencing strategies. Functional analysis and comparative studies are also envisaged in these initiatives, therefore providing basic knowledge of the structures and functions of conifer genomes. The combination of this knowledge with ecophysiological and metabolomic information, among other things, will allow us to start developing a system-level comprehensive understanding of the pathways and processes that lead to changes from the single cell through to conifer individuals and even populations during tree development and under changing environmental conditions.

One of the challenges will be to develop appropriate bioinformatic tools to gradually acquire a system view of plants that integrates multiple types of data across several scales of biologic analysis, with the final long-term goal of the development of predictive models to be applied in both breeding and conservation programs of this unique phylum.

5.7 Acknowledgements

The preparation of this chapter was supported by ProCoGen (289841-FP7) UE project and SustainPine (PLE2009-0016) Plant-KBBE project, as well as by PINCOxSEQ (AGL2012-35175) and RootPine (AGL2011-30462) Spanish National Projects. The authors are very grateful to Susana Ferrándiz for helping with the preparation of the manuscript.

5.8 References

Abarca D, Diaz-Sala C (2009a) Adventitious root formation in conifers. In: Niemi K and Scagel C (eds) *Adventitious Root Formation of Forest Trees and*

Horticultural Plants: From Genes to Applications. Research Signpost, Kerala. pp. 227–57.

Abarca D, Diaz-Sala C (2009b) Reprogramming adult cells during organ regeneration in forest species. *Plant Signaling & Behavior* 4: 793–5.

Abbott E, Hall D, Hamberger B, Bohlmann J (2010) Laser microdissection of conifer stem tissues: Isolation and analysis of high quality RNA, terpene synthase enzyme activity and terpenoid metabolites from resin ducts and cambial zone tissue of white spruce (*Picea glauca*). *BMC Plant Biology* 10: 106.

Ahuja MR, Neale DB (2005) Evolution of genome size in conifers. *Silvae Genetica* 3: 126–37.

Alonso P, Cortizo M, Cantón FR, Fernández B, Rodríguez A, et al. (2007) Identification of genes differentially expressed during adventitious shoot induction in *Pinus pinea* cotyledons by subtractive hybridization and quantitative PCR. *Tree Physiology* 27: 1721–30.

Aranda I, Alía R, Ortega U, Dantas AK, Majada J (2010) Intra-specific variability in biomass partitioning and carbon isotopic discrimination under moderate drought stress in seedlings from four *Pinus pinaster* populations. *Tree Genetics and Genomics* 6: 169–78.

Arrabal C, García-Vallejo MC, Cadahia E, Cortijo M, Fernández de Simon B (2012) Characterization of two chemotypes of Pinus pinaster by their terpene and acid patterns in needles. *Plant Systematics and Evolution* 298: 511–22.

Asante DKA, Yakovlev IA, Fossdal CG, Timmerhaus G, Partanen J, et al. (2009) Effect of bud burst forcing on transcript expression of selected genes in needles of Norway spruce during autumn. *Plant Physiology and Biochemistry* 47: 681–9.

Becker P, Tyree MT, Tsuda M (1999) Hydraulic conductances of angiosperms versus conifers: similar transport efficiency at the whole-plant level. *Tree Physiology* 19: 445–52.

Bedon F, Bomal C, Caron S, Levasseur C, Boyle B, et al. (2010) Subgroup 4 R2R3-MYBs in conifer trees: gene family expansion and contribution to the isoprenoid- and flavonoid oriented responses. *The Journal of Experimental Botany* 61: 3847–64.

Bernstein E, Caudy AA, Hammond SM, Hannon GJ (2001) Role for a bidentate ribonuclease in the initiation step of RNA interference. *Nature* 409: 363–6.

Bowe LM, Coat G, de Pamphilis CW (2000) Phylogeny of seed plants based on all three genomic compartments: extant gymnosperms are monophyletic and Gnetales' closest relatives are conifers. *Proceedings of the National Academy of Sciences USA* 97: 4092–7.

Brinker M, van Zyl L, Liu W, Craig D, Sederoff RR, et al. (2004) Microarray analyses of gene expression during adventitious root development in *Pinus contorta*. *Plant Physiology* 135: 1526–39.

Brodribb TJ, Field TS, Sack L (2010) Viewing leaf structure and evolution from a hydraulic perspective. *Functional Plant Biology* 37: 488–98.

Brodribb TJ, Holbrook NM, Zwieniecki MA, Palma B (2005) Leaf hydraulic capacity in ferns, conifers and angiosperms: impacts on photosynthetic maxima. *New Phytologist* 165: 839–46.

Brodribb TJ, Pittermann J, Coomes DA (2012) Elegance versus speed: examining the competition between conifer and angiosperm trees. *International Journal of Plant Science* 173: 673–94.

Brown GR, Kadel EE, Bassoni DL, Kiehne KL, Temesgen B, et al. (2001) Anchored reference loci in loblolly pine (*Pinus taeda* L.) for integrating pine genomics. *Genetics* 159: 799–809.

Burleigh JG, Barbazuk WB, Davis JM, Morse AM, Soltis PS (2012) Exploring diversification and genome size evolution in extant Gymnosperms through phylogenetic synthesis. *Journal of Botany* (2012) doi:10.1155/2012/292857.

Buschiazzo E, Ritland C, Bohlmann J, Ritland K (2012) Slow but not low: genomic comparisons reveal slower evolutionary rate and higher dN/dS in conifers compared to angiosperms. *BMC Evolutionary Biology* 12: 8.

Cairney J, Zheng L, Cowels A, Hsiao J, Zismann V, et al. (2006) Expressed sequence tags from loblolly pine embryos reveal similarities with angiosperm embryogenesis. *Plant Molecular Biology* 62: 485–501.

Canales J, Flores-Monterrosso A, Rueda-Lopez M, Avila C, Cánovas FM (2010) Identification of genes regulated by ammonium availability in the roots of maritime pine trees. *Amino Acids* 39: 991–1001.

Canales J, Rueda-López M, Craven-Bartle B, Avila C, Cánovas FM (2012) Novel insights into regulation of asparagine synthetase in conifers. *Frontiers in Plant Science* 3: 1–15.

Cannac M, Ferrat L, Bardoni T, Chiaramonti N, Morandini F, et al. (2011) Identification of flavonoids in *Pinus laricio* needles and changes occurring after prescribed burning. *Chemoecology* 21: 9–17.

Chagné D, Brown GR, Lalanne C, Madur D, Pot D, et al. (2003) Comparative genome and QTL mapping between maritime and loblolly pines. *Molecular Breeding* 12: 185–95. doi: 10.1023/A:1026318327911.

Chen F, Tholl D, Bohlmann J, Pichersky E (2011) The family of terpene synthases in plants: a mid-size family of genes for specialized metabolism that is highly diversified throughout the kingdom. *The Plant Journal* 66: 212–29.

Cochard H, Höltta T, Herbette S, Delzon S, Mencuccini M (2009) New insights into the mechanisms of water-stress-induced cavitation in conifers. *Plant Physiology* 151: 949–54.

Côté CL, Boileau F, Roy V, Ouellet M, Levasseur C, et al. (2010) Gene family structure, expression and functional analysis of HD-Zip III genes in angiosperm and gymnosperm forest trees. *BMC Plant Biology* 10: 273.

Crepet WL, Niklas KJ (2009) Darwin's second "abominable mystery": Why are there so many angiosperm species? *American Journal of Botany* 96: 366–81.

de Miguel M, Sánchez-Gómez D, Cervera MT, Aranda I (2012) Functional and genetic characterization of gas exchange and intrinsic water use efficiency in a full-sib family of *Pinus pinaster* Ait. in response to drought. *Tree Physiology* 32: 94–103.

Devey ME, Sewell MM, Uren TL, Neale DB (1999) Comparative mapping in loblolly and radiata pine using RFLP and microsatellite markers. *Theoretical and Applied Genetics* 99: 656–62.

Dolgosheina EV, Morin RD, Aksay G, Sahinalp SC, Magrini V, et al. (2008) Conifers have a unique small RNA silencing signature. *RNA* 14: 1508–15.

Ducrey M, Huc R, Ladjal M, Guehl JM (2008) Variability in growth, carbon isotope composition, leaf gas exchange and hydraulic traits in the eastern Mediterranean cedars *Cedrus libani* and *C. brevifolia*. Tree Physiology 28: 689–701.

Elsik CG, Williams CG (2000) Retroelements contribute to the excess low-copy-number DNA in pine. *Molecular and General Genetics* 264: 47–55.

Elsohly HN, Croom ED, Kopycki WJ, Joshi AS, Elsohly MA, et al. (1995) Concentrations of taxol and related taxanes in the needles of different *Taxus* cultivars. *Phytochemical Analysis* 6: 149–56.

Flanagan LB, Johnsen KH (1995) Genetic variation in carbon isotope discrimination and its relationship to growth under field conditions in full-sib families of *Picea mariana*. *Canadian Journal of Forest Research* 25: 39–47.

Friedmann M, Ralph SG, Aeschliman D, Zhuang J, Ritland K, et al. (2007) Microarray gene expression profiling of developmental transitions in Sitka spruce (*Picea sitchensis*) apical shoots. *Journal of Experimental Botany* 58: 593–614.

Fulton TM, Van der Hoeven R, Eannetta NT, Tanksley SD (2002) Identification, analysis, and utilization of conserved ortholog set markers for comparative genomics in higher plants. *The Plant Cell* 14: 1457–67.

Futamura N, Totoki Y, Toyoda A, Igasaki T, Nanjo T, et al. (2008) Characterization of expressed sequence tags from a full-length enriched cDNA library of *Cryptomeri japonica* male stroboli. *BMC Genomics* 9: 383.

Garcia-Gil MR (2008) Evolutionary aspects of functional and pseudogene members of the phytochrome gene family in Scots pine. *Journal of Molecular Evolution* 67: 222–32.

Gernandt DS, Willyard A, Syring JV, Liston A (2011) The conifers (Pinophyta) In: Plomion C, Bousquet J and Kole C (eds) *Genetics, Genomics and Breeding of Conifers*. Science Publishers, Edenbridge. pp. 1–39.

Hamberger B, Ohnishi T, Hamberger B, Séguin A, Bohlmann J (2011) Evolution of diterpene metabolism: Sitka spruce CYP720B4 catalyzes multiple oxidations in resin acid biosynthesis of conifer defense against insects. *Plant Physiology* 157: 1677–95.

Hammerbacher A, Ralph SG, Bohlmann J, Fenning TN, Gershenzon J, et al. (2011) Biosynthesis of the major tetrahydroxystilbenes in spruce, astringin and isorhapontin, proceeds via resveratrol and is enhanced by fungal infection. *Plant Physiology* 157: 876–90.

Harris PJ (2005) Diversity in plant cell walls. In: RJ Henry (ed.) *Plant Diversity and Evolution: Genotypic and Phenotypic Variation in Higher Plants*. CABI, Wallingford, UK. pp. 201–27.

Hirao T, Watanabe A, Kurita M, Kondo T, Takata K (2008) Complete nucleotide sequence of the *Cryptomeria japonica* D. Don. chloroplast genome and comparative chloroplast genomics: diversified genomic structure of coniferous species. *BMC Plant Biology* 8: 70.

Holmbom B, Eckerman C, Eklund P, Hemming J, Nisula L, et al. (2003) Knots in trees – A new rich source of lignans. *Phytochemistry Reviews* 2: 331–40.

Jennewein S, Long RM, Williams RM, Croteau R (2004) Cytochrome p450 taxadiene 5alpha-hydroxylase, a mechanistically unusual monooxygenase catalyzing the first oxygenation step of taxol biosynthesis. *Chemistry & Biology* 11: 379–87.

Jermstad KD, Eckert AJ, Wegrzyn JL, Delfino-Mix A, Davis DA, et al. (2010) Comparative mapping in *Pinus*: sugar pine (*Pinus lambertiana* Dougl.) and loblolly pine (*Pinus taeda* L.). *Tree Genetics and Genomes* 7: 457–68.

Johnsen KH, Flanagan LB, Huber DA, Major JE (1999) Genetic variation in growth, carbon isotope discrimination, and foliar N concentration in *Picea mariana*: analyses from a half-diallel mating design using field-grown trees. *Canadian Journal of Forest Research* 29: 1727–35.

Joosen RVL, Lammers M, Balk PA, Bronnum P, Konings MCJM, et al. (2006) Correlating gene expression to physiological parameters and environmental conditions during cold acclimation of *Pinus sylvestris*, identification of molecular markers using cDNA microarrays. *Tree Physiology* 26: 1297–313.

Kaundun SS, Lebreton P, Bailly A (2000) Discrimination and identification of coastal Douglas-fir clones using needle flavonoid fingerprints. *Biochemical Systematics and Ecology* 28: 779–91.

Keeling CI, Bohlmann J (2006a) Diterpene resin acids in conifers. *Phytochemistry* 67: 2415–23.

Keeling CI, Bohlmann J (2006b) Genes, enzymes and chemicals of terpenoid diversity in the constitutive and induced defence of conifers against insects and pathogens. *New Phytologist* 170: 657–75.

Keeling CI, Weisshaar S, Lin RP, Bohlmann J (2008) Functional plasticity of paralogous diterpene synthases involved in conifer defense. *Proceedings of the National Academy of Sciences USA* 105: 1085–90.

Keeling CI, Dullat HK, Yuen M, Ralph SG, Jancsik S, et al. (2010) Identification and functional characterization of monofunctional ent-copalyl diphosphate and ent-kaurene synthases in white spruce reveal different patterns for diterpene synthase evolution for primary and secondary metabolism in gymnosperms. *Plant Physiology* 152: 1197–208.

Keeling CI, Weisshaar S, Ralph SG, Jancsik S, Hamberger B, et al. (2011) Transcriptome mining, functional characterization, and phylogeny of a large terpene synthase gene family in spruce (*Picea* spp.). *BMC Plant Biology* 11: 43.

Kidner CA, Martienssen RA (2005) The developmental role of microRNA in plants. *Current Opinion in Plant Biology* 8: 30–44.

Kirst M, Johnson AF, Baucom C, Ulrich E, Hubbard K, et al. (2003) Apparent homology of expressed genes from wood-forming tissues of loblolly pine (*Pinus taeda* L.) with *Arabidopsis thaliana*. *Proceedings of the National Academy of Sciences USA* 100: 7383–8.

Komulainen P, Brown GR, Mikkonen M, Karhu A, García-Gil MR, et al. (2003) Comparing EST-based genetic maps between *Pinus sylvestris* and *Pinus taeda*. *Theoretical and Applied Genetics* 107: 667–78.

Kovach A, Wegrzyn JL, Parra G, Holt C, Bruening GE, et al. (2010) The *Pinus taeda* genome is characterized by diverse and highly diverged repetitive sequences. *BMC Genomics* 11: 420. doi:10.1186/1471-2164-11-420.

Krutovsky KV, Troggio M, Brown GR, Jermstad KD, Neale DB (2004) Comparative mapping in the Pinaceae. *Genetics* 168: 447–61.

Lambert JB, Heckenbach EA, Wu Y (2010) Characterization of plant exudates by principal-component and cluster analyses with nuclear magnetic resonance variables. *Journal of Natural Products* 73: 1643–8.

Lebreton P (1990) La chimiotaxonomie des Gymnospermes. *Bulletin de la Société Botanique de France* 137: 35–46.

Leitch AR, Leitch IJ (2012) Ecological and genetic factors linked to contrasting genome dynamics in seed plants. *The New Phytologist* 194: 629–46.

Lenka SK, Boutaoui N, Paulose B, Vongpaseuth K, Normanly J, et al. (2012) Identification and expression analysis of methyl jasmonate responsive ESTs in paclitaxel producing *Taxus cuspidata* suspension culture cells. *BMC Genomics* 13: 148.

Li X, Wu HX, Southerton SG (2010) Comparative genomics reveals conservative evolution of the xylem transcriptome in vascular plants. *BMC Evolutionary Biology* 10: 190.

Liang C, Wang G, Liu L, Ji G, Fang L, et al. (2007) Conifer EST: an integrated bioinformatics system for data reprocessing and mining of conifer expressed sequence tags (ESTs). *BMC Genomics* 8: 134.

Liewlaksaneeyanawin C, Zhuang J, Tang M, Farzaneh N, Lueng G, et al. (2009) Identification of COS markers in the Pinaceae. *Tree Genetics and Genomes* 5: 247–55.

Lin CP, Huang JP, Wu CS, Hsu CY, Chaw SM (2010) Comparative chloroplast genomics reveals the evolution of pinaceae genera and subfamilies. *Genome Biology and Evolution* 2: 504–17.

Liu W, Thummasuwan S, Sehgal SK, Chouvarine P, Peterson DG (2011) Characterization of the genome of bald cypress. *BMC Genomics* 12: 553.

Loehle C (1988) Tree life history strategies: the role of defenses. *Canadian Journal of Forest Research* 18: 209–22.

Lorenz WW, Sun F, Liang C, Kolychev D, Wang HM, et al. (2006) Water stress-responsive genes in loblolly pine (*Pinus taeda*) roots identified by analyses of expressed sequence tag libraries. *Tree Physiology* 26: 1–16.

Lorenz WW, Alba R, Yu Y-S, Bordeaux JM, Simões M, et al. (2011) Microarray analysis and scale-free gene networks identify candidate regulators in drought-stressed roots of loblolly pine (*P. taeda* L.). *BMC Genomics* 12: 264.

Lorenz WW, Ayyampalayam S, Bordeaux JM, Howe GT, Jermstad KD, et al. (2012) Conifer DBMagic: a database housing multiple de novo transcriptome assemblies for 12 diverse conifer species. *Tree Genetics and Genomes*. doi 10.1007/s11295-012-0547-y.

Lu S, Ying-Hsuan S, Chiang VL (2008) Stress-responsive microRNAs in Populus. *The Plant Journal* 55: 131–51.

Mackay JJ, Dean JFD (2011) Transcriptomics. In: Plomion C, Bousquet J, Kole C (eds) *Genetics, Genomics and Breeding of Conifer Trees*. Science Publishers, Edenbridge. pp. 323–57.

Mackay J, Dean J, Plomion C, Peterson DG, Cánovas F, et al. (2012) Towards decoding the conifer giga-genome. *Plant Molecular Biology* 80: 555–69. doi: 10.1007/s11103-012-9961-7.

Magbanua ZV, Ozkan S, Bartlett BD, Chouvarine P, Saski CA, et al. (2011) Adventures in the enormous: a 1.8 million clone BAC library for the 21.7 Gb genome of loblolly pine. *PLoS ONE* 6: e16214.

Maherali H, Pockman WT, Jackson RB (2004) Adaptive variation in the vulnerability of woody plants to xylem cavitation. *Ecology* 85: 2184–99.

Martin D, Tholl D, Gershenzon J, Bohlmann J (2002) Methyl jasmonate induces traumatic resin ducts, terpenoid resin biosynthesis, and terpenoid accumulation in developing xylem of Norway spruce stems. *Plant Physiology* 129: 1003–18.

Martin DM, Fáldt J, Bohlmann J (2004) Functional characterization of nine Norway spruce TPS genes and evolution of gymnosperm terpene synthases of the TPS-d subfamily. *Plant Physiology* 135: 1908–27.

Martz FO, Peltola R, Fontanay S, Duval RE, Julkunen-Tiito R, et al. (2009) Effect of latitude and altitude on the terpenoid and soluble phenolic composition of juniper (*Juniperus communis*) needles and evaluation of their antibacterial activity in the boreal zone. *Journal of Agricultural and Food Chemistry* 57: 9575–84.

Masterson J (1994) Stomatal size in fossil plants: evidence for polyploidy in majority of angiosperms. *Science* 264: 421–4.

Morgante M, De Paoli E (2011) Toward the conifer genome sequence. In: Plomion C, Bousquet J, Kole C (eds) *Genetics, Genomics and Breeding of Conifers*. Science Publishers, Edenbridge. pp. 389–403.

Morse AM, Peterson DG, Islam-Faridi MN, Smith KE, Magbanua Z, et al. (2009) Evolution of genome size and complexity in *Pinus*. *PLoS ONE* 4: e4332. doi: 10.1371/journal.pone.0004332

Murray BG, Leitch IJ, Bennett MD (2010) Gymnosperm DNA C-values database (release 4.0, Dec. 2010) http://www.kew.org/cvalues.

Neale DB, Sederoff RR (1989) Paternal inheritance of chloroplast DNA and maternal inheritance of mitochondrial DNA in loblolly pine. *Theoretical and Applied Genetics* 77: 212–16.

Nelson D, Werck-Reichhart D (2011) A P450-centric view of plant evolution. *The Plant Journal* 66: 194–211.

Nguyen-Queyrens A, Bouchet-Lannat F (2003) Osmotic adjustment in three-year-old seedlings of five provenances of maritime pine (*Pinus pinaster*) in response to drought. *Tree Physiology* 23: 397–404.

Nilson SE, Assmann SM (2010) Heterotrimeric G proteins regulate reproductive trait plasticity in response to water availability. *New Phytologist* 185: 734–46.

Nilsson L, Carlsbecker A, Sundås-Larsson A, Vahala T (2007) *APETALA2* like genes from *Picea abies* show functional similarities to their Arabidopsis homologues. *Planta* 225: 589–602.

Otto SP, Whitton J (2000) Polyploid incidence and evolution. *Annual Review of Genetics* 34: 401–37.

Packer L, Rimbach G, Virgili F (1999) Antioxidant activity and biological properties of a procyanidin rich extract from Pine (*Pinus maritima*). *Free Radical Biology and Medicine* 27: 704–24.

Paiva JAP, Garnier-Géré PH, Rodrigues JC, Alves A, Santos S, et al. (2008) Plasticity of maritime pine (*Pinus pinaster*) wood-forming tissues during a growing season. *New Phytologist* 179: 1080–94.

Parchman TL, Geist KS, Grahnen JA, Benkman CW, Buerkle CA (2010) Transcriptome sequencing in an ecologically important tree species: assembly, annotation, and marker discovery. *BMC Genomics* 11: 180.

Pavy N, Paule C, Parsons L, Crow JA, Morency M-J, et al. (2005) Generation, annotation, analysis and database integration of 16,500 white spruce EST clusters. *BMC Genomics* 6: 144.

Pelgas B, Bousquet J, Beauseigle S, Isabel N (2005) A composite linkage map from two crosses for the species complex *Picea mariana* x *Picea rubens* and analysis of synteny with other Pinaceae. *Theoretical and Applied Genetics* 111: 1466–88.

Pelgas B, Beauseigle S, Acheré V, Jeandroz S, Bousquet J, et al. (2006) Comparative genome mapping among *Picea glauca*, *P. mariana* x *P. rubens* and *P. abies*, and correspondence with other Pinaceae. *Theoretical and Applied Genetics* 113: 1371–93.

Perdiguero P, Collada C, Barbero MC, García Casado G, Cervera MT, et al. (2012) Identification of water stress genes in *Pinus pinaster* Ait. by controlled progressive stress and suppression-subtractive hybridization. *Plant Physiology and Biochemistry* 50: 44–53.

Peters RJ, Flory JE, Jetter R, Ravn MM, Lee HJ, et al. (2000) Abietadiene synthase from grand fir (*Abies grandis*): characterization and mechanism of action of the 'pseudomature' recombinant enzyme. *Biochemistry* 39: 15 592–602.

Phillips MA, Croteau RB (1999) Resin-based defenses in conifers. *Trends in Plant Science* 4: 184–90.

Pietarinen SP, Willfor SM, Ahotupa MO, Hemmning IE, Holmbom BR (2006) Knotwood and bark extracts: strong antioxidants from waste materials. *Journal of Wood Science* 52: 436–44.

Pittermann J, Sperry JS, Hacke UG, Wheeler JK, Sikkema EH (2005) The torus-margo pit valve makes conifers hydraulically competitive with angiosperms. *Science* 310: 1924.

Pittermann J, Sperry JS, Hacke UG, Wheeler JK, Sikkema EH (2006) Mechanical reinforcement against tracheid implosion compromises the hydraulic efficiency of conifer xylem. *Plant, Cell and Environment* 29: 1618–28.

Pittermann J, Choat B, Jansen S, Stuart SA, Lynn L, et al. (2010) The relationships between xylem safety and hydraulic efficiency in the Cupressaceae: the evolution of pit membrane form and function. *Plant Physiology* 153: 1919–31.

Porth I, Hamberger B, White R, Ritland K (2011) Defense mechanisms against herbivory in *Picea*: sequence evolution and expression regulation of gene family members in the phenylpropanoid pathway. *BMC Genomics* 12: 608.

Prunier J, Laroche J, Beaulieu J, Bousquet J (2011) Scanning the genome for gene SNPs related to climate adaptation and estimating selection at the molecular level in boreal black spruce. *Molecular Ecology* 20: 1702–16.

Ralph SG, Hudgins JW, Jancsik S, Franceschi VR, Bohlmann J (2007a) Aminocyclopropane carboxylic acid synthase is a regulated step in ethylene-dependent induced conifer defense. Full-length cDNA cloning of a multigene family, differential constitutive, and wound- and insect-induced expression, and cellular and subcellular localization in spruce and douglas fir. *Plant Physiology* 143: 410–24.

Ralph SG, Jancsik S, Bohlmann J (2007b) Dirigent proteins in conifer defense II: Extended gene discovery, phylogeny, and constitutive and stress-induced gene expression in spruce (*Picea* spp.). *Phytochemistry* 68: 1975–91.

Ralph SG, Chun HJE, Kolosova N, Cooper D, Oddy C, et al. (2008) A conifer genomics resource of 200,000 spruce (*Picea*s pp.) ESTs and 6,464 high-quality, sequence-finished full-length cDNAs for Sitka spruce (*Picea sitchensis*). *BMC Genomics* 9: 484.

Rigault P, Boyle B, Lepage P, Cooke JEK, Bousquet J, et al. (2011) A white spruce gene catalog for conifer genome analyses. *Plant Physiology* 157: 14–28.

Ritland K (2012) Genomics of a phylum distant from flowering plants: conifers. *Tree Genetics & Genomes* 8: 573–82.

Ritland K, Krutovsky KV, Tsumura Y, Pelgas B, Isabel N, et al. (2011) Genetic mapping in conifers. In: Plomion C, Bousquet J, Kole C (eds) *Genetics, Genomics and Breeding in Conifers*. Science Publishers, Edenbridge. pp. 196–238.

Rontein D, Onillon S, Herbette G, Lesot A, Werck-Reichhart D, et al. (2008) CYP725A4 from yew catalyzes complex structural rearrangement of taxa-4(5),11(12)-diene into the cyclic ether 5(12)-oxa-3(11)-cyclotaxane. *Journal of Biological Chemistry* 283: 6067–75.

Sánchez C, Vielba JM, Ferro E, Covelo G, Solé A, et al. (2007) Two *SCARECROW-LIKE* genes are induced in response to exogenous auxin in rooting-competent cuttings of distantly related forest species. *Tree Physiology* 27: 1459–70.

Sánchez-Gómez D, Velasco-Conde T, Cano-Martín FJ, Guevara MA, Cervera MT, et al. (2011) Inter-clonal variation in functional traits in response to drought for a genetically homogeneous Mediterranean conifer. *Environmental and Experimental Botany* 70: 104–9.

Sedel'nikova TS, Muratova EN, Pimenov AV (2011) Variability of chromosome numbers in gymnosperms. *Biology Bulletin Reviews* 1: 100–9.

Shepherd M, Williams CG (2008) Comparative mapping among subsection Australes (genus *Pinus*, family Pinaceae). *Genome* 51: 320–31.

Slimestad R (2003) Flavonoids in buds and young needles of *Picea*, *Pinus* and *Abies*. *Biochemical Systematics and Ecology* 31: 1247–55.

Smeds AI, Eklund PC, Sjöholm RE, Willför SM, Nishibe S, et al. (2007) Quantification of a broad spectrum of lignans in cereals, oilseeds, and nuts. *Journal of Agricultural and Food Chemistry* 55: 1337–46.

Solé A, Sánchez C, Vielba JM, Valladares S, Abarca D, et al. (2008) Characterization and expression of a *Pinus radiata* putative ortholog to the *Arabidopsis SHORT-ROOT* gene. *Tree Physiology* 28: 1629–39.

Sperry JS, Sullivan JEM (1992) Xylem embolism in response to freeze-thaw cycles and water stress in ring-porous, diffuse-porous, and conifer species. *Plant Physiology* 100: 605–13.

Sperry JS, Hacke UG, Pittermann J (2006) Size and function in conifer tracheids and angiosperm vessels. *American Journal of Botany* 93: 1490–500.

Springer NM, Xu X, Barbazuk WB (2004) Utility of different gene enrichment approaches toward identifying and sequencing the maize gene space. *Plant Physiology* 136: 3023–33.

Stevanovic T, Diout PN, García-Pérez ME (2009) Bioactive polyphenols from healthy diets and forest biomass. *Current Nutrition & Food Science* 5: 264–95.

Strauss SH, Palmer JD, Howe GT, Doerksen AH (1988) Chloroplast genomes of two conifers lack a large inverted repeat and are extensively rearranged. *Proceedings of the National Academy of Sciences USA* 85: 3898–902.

Sun Y-H, Shi R, Zhang X-H, Chiang VL, Sederoff RR (2011) MicroRNAs in trees. *Plant Molecular Biology* 80: 37–53.

Tappert R, Wolfe AP, McKellar RC, Tappert MC, Muehlenbachs K (2011) Characterizing modern and fossil gymnosperm exudates using micro-Fourier transform infrared spectroscopy. *International Journal of Plant Sciences* 172: 120–38.

Verne S, Jaquish B, White R, Ritland C, Ritland K (2011) Global transcriptome analysis of constitutive resistance to the *White Pine* weevil in spruce. *Genome Biology and Evolution* 3: 851–67.

Vielba JM, Díaz-Sala C, Ferro E, Rico S, Lamprecht M, et al. (2011) *CsSCL1* is differentially regulated upon maturation in chestnut microshoots and is specifically expressed in rooting-competent cells. *Tree Physiology* 31: 1152–60.

Wakasugi T, Tsudzuki J, Ito S, Nakashima K, Tsudzuki T, et al. (1994) Loss of all NDH genes as determined by sequencing the entire chloroplast genome of the black pine *Pinus thunbergii*. *Proceedings of the National Academy of Sciences USA* 91: 9794–8.

Wildung MR, Croteau R (1996) A cDNA clone for taxadiene synthase, the diterpene cyclase that catalyzes the committed step of taxol biosynthesis. *Journal of Biological Chemistry* 271: 9201–4.

Willför SM, Ahotupa MO, Hemmning IE, Reunanen MHT, Eklund PC, et al. (2003) Antioxidant activity of knotwood extractives and phenolic compounds of selected tree species. *Journal of Agricultural and Food Chemistry* 51: 7600–6.

Willför SM, Ali M, Karonen M, Reunanen M, Arfan M, et al. (2009) Extractives in bark of different conifer species growing in Pakistan. *Holzforschung* 63: 551–8.

Willson CJ, Jackson RB (2006) Xylem cavitation caused by drought and freezing stress in four co-occurring Juniperus species. *Physiologia Plantarum* 127: 374–82.

Wu CS, Wang YN, Liu SM, Chaw SM (2007) Chloroplast genome (cpDNA) of *Cycas taitungensis* and 56 cp protein-coding genes of *Gnetum parvifolium*: insights into cpDNA evolution and phylogeny of extant seed plants. *Molecular Biology and Evolution* 24: 1366–79.

Wu CS, Lin CP, Hsu CY, Wang R, Chaw SM (2011) Comparative chloroplast genomes of pinaceae: insights into the mechanism of diversified genomic organization. *Genome Biology and Evolution* 3: 309–19.

Yakovlev II, Fossdal CG, Johnsen Ø (2010) MicroRNAs, epigenetic memory and climatic adaptation in *Norway spruce*. *New Phytologist* 187: 1154–69.

Yang S-H, Loopstra CA (2005) Seasonal variation in gene expression for loblolly pines (*Pinus taeda*) from different geographical regions. *Tree Physiology* 25: 1063–73.

Yesil-Celiktas O, Otto F, Parlar H (2009) A comparative study of flavonoid contents and antioxidant activities of supercritical CO_2 extracted pine barks grown in different regions of Turkey and Germany. *European Food Research and Technology* 229: 1101–5.

Zhang J, Marshall JD (1994) Population differences in water-use efficiency of well-watered and water-stressed western larch seedlings. *Canadian Journal of Forest Research* 24: 92–9.

Zhang J, Marshall JD, Jaquish BC (1993) Genetic differentiation in carbon isotope discrimination and gas exchange in *Pseudotsuga menziesii*. *Oecologia* 93: 80–7.

Zulak KG, Bohlmann J (2010) Terpenoid biosynthesis and specialized vascular cells of conifer defense. *Journal of Integrative Plant Biology* 52: 86–97.

Zwieniecki MA, Boyce CK, Holbrook NM (2004) Functional design space of single-veined leaves: role of tissue hydraulic properties in constraining leaf size and shape. *Annals of Botany* 94: 507–13.

Zwieniecki MA, Stone HA, Leigh A, Boyce CK, Holbrook NM (2006) Hydraulic design of pine needles: one-dimensional optimization for single-vein leaves. *Plant Cell and Environment* 29: 803–9.

6

Cryptochrome genes modulate global transcriptome of tomato

Loredana Lopez and Gaetano Perrotta, ENEA, Trisaia Research Centre, Italy

DOI: 10.1533/9781908818478.97

Abstract: Light is an essential and variable environmental factor that governs plant growth and development. Light signals are perceived by at least four distinct families of photoreceptors: the red (R)/far-red (FR) light sensing phytochromes (PHYs) and the UV-A blue light sensing cryptochromes (CRYs), phototropins and zeitlupes. Day/night and temperature cycles, as well as other cycling environmental parameters, are highly important to most organisms. Light signaling pathways and circadian control are interconnected processes, with photoreceptors involved in entrainment of the clock, and the clock involved in the regulation of photoreceptor genes. In this article a review is presented on the functional studies elucidating the role of cryptochromes in mediating light-regulated gene expression in plants. DNA microarrays and tomato transcriptomic analysis showed that cryptochromes influence the diurnal global transcription profiles in tomato and are involved in hormone–photoreceptor crosstalk.

Key words: tomato, circadian rhythms, photoperiod, cryptochromes, phytohormones, light-regulated gene expression.

6.1 Introduction

Light is an essential and variable environmental factor that governs plant growth and development. In addition to being the main energy source for plants, light also controls multiple developmental processes throughout the plant life cycle, including seed germination, seedling de-etiolation, phototropism, shade avoidance, circadian rhythms, and flowering time (Jiao et al., 2007).

Light signals are perceived by at least four distinct families of photoreceptors: the red (R)/far-red (FR) light-sensing phytochromes (PHYs), the UV-A blue-light-sensing cryptochromes (CRYs), phototropins and zeitlupes (Möglich et al., 2010). Plant PHY and CRY photoreceptors are best studied in *Arabidopsis*. The *Arabidopsis* genome encodes five phytochromes (PHYA–PHYE). PHYA is most abundant in dark-grown seedlings, whereas its level drops rapidly upon exposure to R or white light. In light-grown plants, PHYB is the most abundant phytochrome, while PHYC–PHYE are less abundant (Franklin and Quail, 2010). Three cryptochrome genes, CRY1, CRY2 and CRY3, have been described so far in the *Arabidopsis* genome. CRY1 and CRY2 act primarily in the nucleus (Wu and Spalding, 2007; Yu et al. 2007), whereas CRY3 probably functions in chloroplasts and mitochondria (Kleine et al., 2003).

Two CRY1s (CRY1a and CRY1b), one CRY2 and one CRY3 gene have been identified in tomato (Facella *et al.*, 2006; Perrotta *et al.*, 2000, 2001). In tomato, the role of the CRY1a gene has been elucidated through the use of antisense (Ninu et al., 1999) and mutant (Weller et al., 2001) plants. CRY1a controls seedling photomorphogenesis, anthocyanin accumulation, and adult plant development. No major effects of CRY1a on flowering time or tomato fruit pigmentation have been observed. The overexpression of tomato CRY2 causes phenotypes similar to but distinct from their *Arabidopsis* counterparts (hypocotyls and internode shortening under both low and high fluency blue light), but also several novel ones, including a high pigment phenotype, resulting in overproduction of anthocyanins and chlorophyll in leaves and of flavonoids and lycopene in fruits (Giliberto et al. 2005).

In tomato, phytochromes are encoded by five genes: PHYA, PHYB1, PHYB2, PHYE and PHYF (Hauser et al., 1995). Phylogenetic analyses showed orthology between PHYA, PHYE and PHYC/F gene pairs in *Arabidopsis* and tomato; tomato PHYB1 and PHYB2 were originated by an independent duplication (Pratt et al., 1995). PHYA and PHYB1 are involved in the mediation of de-etiolation responses to R light (Van Tuinen et al., 1995a, b). Although the phyAphyB1 double mutant is blind to low R light irradiance, it de-etiolated normally under white light. The phenotype of phyAphyB1phyB2 mutants under natural daylight indicated an important role for PHYB2 in this residual response (Kerckhoffs et al., 1999), showing that PHYB2 is also active in R-sensing (Weller et al., 2000).

Photoreceptors sense and transduce light signals through distinct intracellular signaling pathways to generate a wide range of responses. Most of these responses are triggered by modulating the expression of hundreds of light-regulated genes, which ultimately lead to adaptive changes at the cellular and systemic levels.

Day/night and temperature cycles, as well as other cycling environmental parameters, are highly important to most organisms. To follow and anticipate these cycles, plants possess an internal timekeeping system, the circadian clock, which runs on a period of approximately 24 h and regulates several molecular and physiological responses (Harmer, 2009; McWatters and Devlin, 2011). Circadian clocks are based on one or more interlocking transcriptional feedback loops between a set of key genes.

Our current understanding of the plant circadian clock derives mostly from genetic studies in *Arabidopsis thaliana* and rice (Hayama and Coupland, 2004). Commonly, the circadian clock system is divided into three parts (Dunlap, 1999): an input pathway that entrains the clock by transmitting light or temperature signals to the core oscillator, the central oscillator (the clock), responsible for driving 24-h rhythms, and the output signals that generate the fluctuation of a wide range of molecular, biochemical and developmental responses. The molecular components of the *Arabidopsis* circadian system comprise at least three interlocking transcriptional/translational feedback loops (Pruneda-Paz et al., 2009). The first identified loop (termed 'the central oscillator') contains two light-regulated Myb domain transcription factors, CIRCADIAN AND CLOCK ASSOCIATED 1 (CCA1) and LONG HYPOCOTYL (LHY), which, together, regulate abundance of the PSEUDO RESPONSE REGULATOR (PRR) protein, TIMING OF CAB1 (TOC1) and the recently identified CCA1 Hiking Expedition (CHE), a TCP transcription factor. A second interlocking loop incorporates the PRR proteins PRR7 and PRR9, while a third loop involves Gigantea (GI) and possibly PRR5 (McClung, 2010). Rather than being separate processes, light signaling pathways and circadian control are interconnected, with photoreceptors involved in entrainment of the clock, and the clock involved in the regulation of photoreceptor genes (Fankhauser and Staiger, 2002).

6.2 Cryptochrome functions

Plants depend on CRYs and other photoreceptors to sense environmental cues, such as irradiance, day–night transition, photoperiods, and light quality for optimal growth and development. It is well known that *Arabidopsis* CRY1 and CRY2 mediate primarily blue-light regulation of de-etiolation and photoperiodic control of flowering, respectively (Ahmad and Cashmore, 1993; Guo et al., 1998). In addition, these two photoreceptors regulate other aspects of plant growth and development, including entrainment of the circadian clock (Somers et al., 1998; Yanovsky et al., 2001), guard cell development and stomatal opening (Kang et al., 2009; Mao et al., 2005), root growth (Zeng et al., 2010), plant height (Giliberto et al., 2005; Weller et al., 2001), fruit and ovule size (El-Assal et al., 2004), tropic growth (Tsuchida-Mayama et al., 2010), apical dominance (Giliberto et al., 2005; Weller et al., 2001), apical meristem activity (Lopez-Juez et al., 2008), programmed cell death (Danon et al., 2006), the high-light stress response (Kleine et al., 2007), osmotic stress response (Xu et al., 2009), shade avoidance (Keller et al., 2011) and responses to bacterial and viral pathogens (Jeong et al., 2010; Wu and Yang, 2010). *Arabidopsis* CRY3 belongs to the CRY-DASH clade of the photolyase/CRY superfamily. Cry-DASH proteins bind both double-stranded and single-stranded DNA and can also repair single-stranded DNA containing cyclobutane-pyrimidine-dimer (CPD) lesions, but their function in signal transduction, if any, is not yet known (Huang et al., 2006; Pokorny et al., 2008).

6.3 Role of cryptochromes in mediating light-regulated gene expression in plants

In the last few years, extensive progress has been made in elucidating the molecular, cellular, and biochemical mechanisms underlying light perception and subsequent signal transduction (Jenkins, 2009; Li et al., 2011; Pedmale et al., 2010; Yu et al., 2010).

Day/night fluctuations in the quantity and quality of illumination alter the plant's energy and carbon resources. In daylight, photosynthesis results in carbon fixation and starch accumulation that supports metabolic and growth activities in the dark. The alteration between light and darkness, circadian regulation, and the resultant fluctuation of sugars induce changes in the accumulation and stability of mRNA transcripts, as shown in *Arabidopsis thaliana* (Blasing et al., 2005; Graf et al., 2010; Lidder et al., 2005). Traditional genetic and molecular approaches have been powerful in identifying various key regulators and their positions within these signaling cascades. In most instances, the roles of individual photoreceptors are studied in the context of specific responses and/or developmental stages. By using traditional approaches, plant biologists had identified more than 100 individual genes whose expression is regulated by light (Fankhauser and Chory, 1997; Kuno and Furuya, 2000). However, genomic studies conducted in recent years have greatly expanded on these traditional approaches by providing an overall picture of the genome-wide changes that occur during photomorphogenesis, that is, the process through which light controls several developmental responses, such as chloroplast differentiation, chlorophyll accumulation, leaf expansion, and inhibition of stem and internode elongation. The introduction of microarray technology allowed light-regulated gene expression to be studied at increasingly larger scales, culminating in genome-wide analysis.

Lately, the introduction of high-throughput next-generation sequencing technologies has further revolutionized transcriptomics by allowing RNA analysis through massive cDNA sequencing (RNA-seq) (Ozsolak and Milos, 2011) to examine light regulation of plant transcriptomes (Zhang et al., 2011).

A genomic study conducted in both *Arabidopsis* and rice and a number of Solanaceae demonstrated that light induces massive reprogramming of the transcriptome: at least 30% of genes are regulated by white light (Jiao et al., 2005; Ma et al., 2001; Rutitzky et al., 2009). In addition, organ-specific expression profiles during seedling photomorphogenesis indicate that light effects diverge significantly in separate organs. Furthermore, different colors of light, even though they are perceived and transduced by distinct photoreceptors, largely affect the expression of the same fraction of the genome (Ma et al., 2001; Peschke and Kretsch, 2011). Analysis of these light-regulated genes revealed more than 26 cellular pathways that are coordinately regulated by light (Jiao et al., 2005; Ma et al., 2001).

Significant progress has been made over the past few years in understanding the photoexcitation and signal transduction mechanisms of *Arabidopsis* CRYs (Chaves et al., 2011; Liu H et al., 2011; Yu et al., 2010). The photoexcited CRY molecules undergo several biophysical and biochemical changes, including electron transfer, phosphorylation and ubiquitination to propagate light signals. The signal transduction mechanism of *Arabidopsis* cryptochromes remains not fully understood, although it is generally agreed that cryptochromes interact with signaling proteins to alter gene expression at both transcriptional and post-translational levels and consequently the metabolic and developmental programs of plants (Jiao et al., 2007; Lin and Shalitin, 2003; Partch et al., 2005). Approximately 5–25% of genes in the *Arabidopsis* genome change their expression in response to blue light; most of the changes are mediated by CRY1 and CRY2 (Ma et al., 2001; Ohgishi et al., 2004; Sellaro et al., 2009).

The expression of many CRY-regulated genes is also regulated by other signaling pathways, such as phytochromes and phytohormones, suggesting that the CRY-dependent photomorphogenesis is intimately integrated with the general regulatory networks that control plant development. *Arabidopsis* CRYs mediate blue-light control of gene expression via at least two mechanisms: light-dependent modulation of transcription via bHLH (basic-helix-loop-helix) transcription factor CIB1 (CRY–CIBs pathway) (Liu et al., 2008) and light-dependent suppression of proteolysis (CRY–SPA1/COP1 pathway) (Zuo et al., 2011). Both mechanisms are involved with blue-light-dependent protein–protein interactions of CRYs and the signaling proteins. CRY-interacting basic-helix-loop-helix 1 (CIB1) is the first blue-light-dependent CRY2-interacting protein identified in *Arabidopsis* (Liu et al., 2008). CIB1 positively regulates floral initiation in a CRY2-dependent manner and it interacts with the chromatin of the promoter DNA of the FLOWERING LOCUS T (FT) gene, which encodes a mobile transcriptional regulator that migrates from leaves to the apical meristem to activate transcription of floral meristem identity genes (Corbesier and Coupland, 2006). Additional results suggested that multiple CIB proteins act redundantly in the CRY2–CIB signal transduction pathway to mediate photoperiodic promotion of floral initiation (Liu et al., 2008). CRYs can also indirectly modulate gene expression via post-transcriptional mechanisms by interacting with the SPA1/COP1 complex (Lian et al., 2011; Liu B et al., 2011; Zuo et al., 2011). It is well known that CRYs mediate blue-light suppression of the E3 ubiquitin ligase COP1 and COP1-dependent proteolysis to affect gene expression (Jiao et al., 2007; Ma et al., 2001). For example, CRY1 mediates blue-light suppression of the COP1-dependent degradation of the bZIP transcription factors LONG HYPOCOTYL5 (HY5), HY5 HOMOLOGUE (HYH), and the bHLH transcription factor Long Hypocotyl in Far-Red 1 (HFR1), which regulate transcription of genes required for the de-etiolation response (Jiao et al., 2007; Lee et al., 2007). Similarly, CRY2 mediates blue-light suppression of the COP1-dependent protein degradation of a major transcriptional regulator of floral initiation, CONSTANS (CO). The CO protein is a critical positive regulator of flowering in long day, which promotes flowering initiation by activating transcription of the florigen gene FT (Searle and Coupland, 2004). It has been shown that COP1 physically interacts with CO in vivo, and that COP1 facilitates ubiquitination of CO in vitro (Jang et al., 2008; Liu et al., 2008). CRYs are required for the accumulation of the CO protein in blue light (Valverde et al., 2004), whereas COP1 promotes CO degradation in the absence of blue light (Jang et al., 2008; Liu et al., 2008). These observations argue that CRYs mediate blue-light-dependent suppression of the COP1 activity to facilitate CO accumulation and floral initiation in response to photoperiodic signals. Additional studies in *Arabidopsis* highlighted the role of suppressor of phytochrome A (SPA1) in the CRY signal transduction; in fact, it was found that *Arabidopsis* CRY1 and CRY2 interact with SPA1 in response to blue-light but not red light, and that SPA1 acts genetically downstream of CRY1 and CRY2 to mediate blue-light suppression of the COP1-dependent degradation of HY5 and CO, respectively. Because SPA1 is known to interact physically with COP1 in a light-dependent manner and it is a positive regulator of COP1 (Lian et al., 2011), the blue-light-dependent CRY–SPA1 interaction appears to solve, at least partially, the puzzle of how CRYs, which showed no light-dependent interaction with COP1 in the previous studies, mediate light-dependent suppression of COP1 (Lian et al., 2011).

Blue-light-induced change of gene expression in both nuclear and plastid genomes is probably the major mechanism underlying development of plastids during de-etiolation (Chen et al., 2004; Jiao et al., 2007; Kasemir, 1979). Cryptochromes appear to mediate blue-light induction of changes in gene expression via both transcriptional and post-transcriptional mechanisms. Cryptochromes mediate blue-light induction of nuclear genes that encode plastid proteins required not only for photosynthesis and other functions of plastids, such as chlorophyll a/b binding (CAB), ribulose-1, 5-bisphosphate carboxylase/oxygenase small subunit (rbcS), and chalcone synthase (CHS) (Fuglevand et al., 1996; Ma et al., 2001), but also for components of the plastid transcriptional apparatus (Ma et al., 2001; Mochizuki et al., 2004; Thum et al., 2001; Tsunoyama et al., 2002). For example, the blue-light-induced transcription of the plastid-encoded *psbA* and *psbD* genes that encode photosystem II (PSII) reaction center protein D1 and D2, respectively, are controlled by one of the six nuclear-encoded sigma factors, SIG5. It has been shown that CRY1 is the major photoreceptor mediating blue-light induction of the expression of the nucleus-encoded *SIG5* gene (Mochizuki et al., 2004; Nagashima et al., 2004; Thum et al., 2001; Tsunoyama et al., 2002). Therefore, cryptochrome regulation of the supply of nuclear-encoded proteins, which are required for the plastid transcription machinery controlled by the plastid-encoded RNA polymerase, may play a critical role in chloroplast development during de-etiolation.

Light signals perceived and transduced by phytochromes and cryptochromes entrain the clock to the external light/dark cycles. The photoreceptors involved in light input to the circadian clock have been elucidated in *Arabidopsis*, through analyses of null mutants. To date, roles have been established for phytochromes A, B, D, and E and the cryptochromes 1 and 2 (Devlin and Kay, 2000; Somers et al., 1998).

Microarray technology has been extensively exploited to survey genome-wide circadian regulation of gene expression, and has provided important insights into clock function (Covington et al., 2008; Harmer et al., 2000; Michael et al., 2008; Schaffer et al., 2001). It has been shown that a large fraction of the plant transcriptome, as much as one-third of all expressed genes, is regulated by the circadian clock (Covington et al., 2008). These results highlighted that the circadian clock controls many metabolic, developmental, and physiological processes in a time-of-day-specific manner in plants (de Montaigu et al., 2010).

It is well known that transcription factors play key roles in the light-regulated transcriptional networks (Tepperman et al., 2001, 2006; Ulm et al., 2004); the rapid responsiveness of transcription factors indicates that they may represent a master set of transcriptional regulators that orchestrate the expression of the downstream target genes in the light-regulated transcriptional networks. However, only a few key transcription factors (such as HY5, PIF1 and FHY3) (Zhang et al., 2011) have been surveyed so far for their genome-wide binding sites. Additional studies are required to investigate the global binding sites of other key transcription factors in light signaling, and to understand how these transcription factors co-regulate their common target genes, and how they interact with each other.

Hormone regulation represents another layer of control for light-regulated gene expression in plant development. The involvement of plant hormones, such as gibberellic acid (GA), in light-regulated development, which includes seed germination and seedling photomorphogenesis, has been known for many years (Neff et al., 2006). GAs can modulate several molecular processes during the plant life cycle, including germination, vegetative growth, and flowering, through transcriptional regulation of target genes (Yamaguchi, 2008).

Photoreceptor–hormone interactions have been reported to regulate a number of light responses; phytochromes and GAs are indeed involved (together with auxins and ethylene) in regulating shade avoidance responses, which maximize light capture by positioning the leaves out of the shade (Vandenbussche et al., 2005). There are several pieces of evidence of interactions between photoreceptors and hormones during plant development. Many studies have suggested that phytochromes and cryptochromes influence the activities of auxin in order to regulate plant growth. Indeed, PHYA, PHYB, and CRY1 promote light-dependent effects of the auxin transport inhibitor 1-N-naphthylphthalamic acid on both hypocotyls and root elongation in *Arabidopsis* (Canamero et al., 2006; Jensen et al., 1998). Other reports indicate that cryptochromes regulate the transcription of AUX/IAA genes (Folta et al., 2003) and that AUX/IAAs are phosphorylated by PHYA (Colon-Carmona et al., 2000). The relationship between cryptochromes and GA in the blue-light responses is clear in *Arabidopsis*. It has been found in pea that CRY1 and PHYA redundantly regulate GA2ox and GA3ox expression and GA signaling (Foo et al., 2006; Symons and Reid, 2003). A recent report demonstrated that cryptochromes mediate blue-light regulation of GA catabolic/metabolic genes, which affect GA levels and hypocotyl elongation (Zhao et al., 2007). Furthermore, Vandenbussche and collegues (Vandenbussche et al., 2007) concluded that HY5, a positive regulator of photomorphogenesis induced by CRY1 and CRY2 (Wang et al., 2001), represents a point of convergence between cryptochrome and cytokinin signaling pathways.

6.4 Cryptochromes influence the diurnal global transcription profiles in tomato

Tomato (*Solanum lycopersicum*) is a major crop plant and a model system for fruit development. A vast collection of mutants and introgression lines is available, and sequencing of its genome has recently been published (Tomato Genome Consortium, 2012).

In order to understand the mechanism of action of cryptochromes in tomato and to define the role of these photoreceptors in mediating light-regulated gene and protein expression, in recent years a number of molecular approaches have contributed to clarifying how cryptochrome-mediated light signals affect the global expression profiles of the tomato genome (Facella et al., 2006, 2008; Lopez et al., 2012).

The use of large-scale microarray-transcriptomic profiling in wild type (*WT*) tomato plants grown under a light cycle of 16 h light/8 h darkness (LD) evidenced that in tomato, as in *Arabidopsis*, diurnal rhythms in gene expression affect a large portion of the transcriptome (Facella et al., 2008; Schaffer et al., 2001). A diurnally regulated expression pattern was seen in 1016 transcripts, corresponding to 15% of all spots included in the array (Facella et al., 2008). The majority of diurnally regulated genes showed an expression peak in the middle of the light phase, while the other transcription peaks appeared evenly distributed at the other time points, supporting the occurrence of highly coordinated and alternated metabolic processes (Facella et al., 2008). Given the cyclic nature of many physiological processes driven by photo- and thermo-oscillations (Michael et al., 2008), it is expected that the majority of transcripts involved in the biosynthesis of mitochondrial and cytosolic proteins peak in the middle of the light phase. This can be attributed to the fact that the biosynthetic processes correlated with photosynthesis and energy metabolism are usually more active in light hours. Similarly, the fact that several transcripts coding for proteins involved in transport, in transferase activity and in the transcription control machinery were also abundant at dusk and during the night indicates that, during the hours of darkness, synthesis of these proteins is still active (Facella et al., 2008). Several transcripts with higher levels during daylight, encoding proteins involved in photosynthesis and a number of transcription factors, such as MYB, WRKY, and bHLH, could have a major role in adapting tomato plants to day conditions, such as excess of light and higher temperatures (Facella et al., 2008). Conversely, several transcripts relatively more abundant in the dark phase are related to biochemical processes occurring in darkness (Facella et al., 2008). Several tomato homologues of the genes involved in the circadian clock feedback-loop in *Arabidopsis* (Hotta et al., 2007) were found to oscillate in a similar phase in tomato (Facella et al., 2008): the morning element LHY was up-regulated at dawn, while the PRR7, thought to establish a negative loop with CCA1/LHY, was more expressed during daylight and down-regulated at dawn (Facella et al., 2008). ELF4 and GI, which are putatively involved in feedback-loops with CCA1/LHY and TOC1/LUX, respectively (Locke et al., 2006), were accordingly more expressed around dusk (Facella et al., 2008). These results suggest that the basic molecular machinery of the circadian clock is conserved in higher plants. Furthermore, the fact that a number of other elements, such as FKF1 (Nelson et al., 2000) and SPA1 (Ishikawa et al., 2006) related to the input/output signaling of the *Arabidopsis* circadian clock, but also involved in flowering and light transduction, showed similar transcript fluctuations in tomato (Facella et al., 2008) suggests that molecular interactions between the clock core and input/output pathways are also partially conserved.

Comparative transcription and proteomic analyses in tomato WT versus a CRY2 transgenic over-expressor (*CRY2-OX*) genotype (Giliberto et al., 2005) under a diurnal rhythm showed that tomato CRY2 profoundly affects both gene and protein expression in response to a daily light cycle (Lopez et al., 2012). It is not surprising that alterations identified in some of the photoreceptor genes result in extensive perturbations of gene expression with profound effects on cell differentiation and biogenesis of organelles such as plastids and possibly mitochondria (Jiao et al., 2007; Kleine et al., 2003). It is conceivable that light modulates the energetic metabolism of tomato through a fine CRY-mediated transcriptional control (Lopez et al., 2012). In a diurnal cycle, as many as 45% of tomato transcripts represented on the microarray are altered between *CRY2-OX* and *WT* genotypes, with an interesting accumulation of differences at dawn and at dusk (Lopez et al., 2012). At the proteome level, differences were instead concentrated in the middle of the light phase and at dusk (Lopez et al., 2012). The data suggest that some cellular pathways are specifically altered by overexpression of CRY2 in tomato. Indeed, a large number of photosynthesis-related genes, including components of the light and dark reactions and of starch and sucrose biosynthetic pathways, were found among the genes differentially expressed in *CRY2-OX* (Lopez et al., 2012). Light regulation of nuclear photosynthesis genes is a well-known phenomenon, controlled by a complex interplay of photoreceptor-, circadian-, redox- and sugar-mediated signaling pathways (Eberhard et al., 2008). Both glycolysis and the tricarboxylic acid cycle that oxidizes photosynthesis-produced carbohydrates, providing energy and a carbon source for many biosynthetic pathways, were altered in *CRY2-OX* (Lopez et al., 2012). Furthermore, overexpression of CRY2 influences secondary metabolism, such as phenylpropanoid, phenolic and flavonoid/anthocyanin biosynthesis pathways (Lopez et al., 2012). One of the most interesting results is that CRY2 overexpression causes a massive phase shift in transcript abundance (Lopez et al., 2012). Interestingly, most of the shifted transcripts are involved in ethylene biosynthesis and signal transduction. Ethylene can regulate more than 400 genes in *Arabidopsis*, including genes for the biosynthesis of hormones, signaling components, transcription factors, structural proteins, photosynthetic and metabolic enzymes, and plant defense-related proteins (Zhu and Guo, 2008). Accordingly, most of the transcripts with a 'phase shift' in tomato assays are involved in biotic and abiotic stress responses (Lopez et al., 2012). It is well known that in *Arabidopsis* CRYs are involved in the measurement of photoperiod length and in the entrainment of the circadian machinery to the light stimulus (Guo et al., 1998; Yanovsky et al., 2001). With respect to these important functions, the most sensitive phases of the day are dawn and dusk, that is, the start and the end of the light period. Then, a working hypothesis could be that an increase in active CRY2 protein can cause an 8-h anticipation of the transcription abundance peak that is usually observed in the middle of the light phase (Lopez et al., 2012). In addition, the observed up-regulation of the photorespiratory-related genes in the CRY2-OX genotype (Lopez et al., 2012) raises the intriguing possibility that in tomato CRY2 is able to operate a multi-organelle control of photorespiration. Photorespiration is the metabolism of phosphoglycolate that is produced during oxygenation catalyzed by the enzyme RubisCO and inhibits photosynthesis by interfering with CO_2 fixation by RubisCO (Linka and Weber, 2005). Given the fact that, in tomato, CRY2 overexpression causes alterations of the circadian expression pattern of several gene transcripts (Facella et al., 2008), it cannot be ruled out that the observed up-regulation of the photorespiratory-related genes is the result of the combined action of CRY 2 and the circadian clock machinery.

A focused approach to a number of photosensory and clock-related genes by a quantitative real-time polymerase chain reaction (QRT-PCR), in both *WT* tomato and genotypes with altered cryptochrome gene expression (tomato mutant deficient in CRY1a signaling (*cry1a-*) and transgenic over-expressor *CRY2-OX* (Facella et al., 2008; Giliberto et al., 2005; Weller et al., 2001), provided further information about transcription alterations induced by cryptochromes through the circadian clock machinery, as well as about regulatory interactions between different photoreceptors (Facella et al., 2008). A large series of transcript oscillations that shed light on the complex network of interactions among tomato photoreceptors and clock-related genes were identified (Facella et al., 2008). Cryptochromes regulate phytochrome transcript levels, resulting in changes in transcript abundance, phase and cycling amplitude. In particular, overexpression of CRY2 had an impact not only on day/night fluctuations but also on rhythmicity under constant light conditions (Facella et al., 2008). Additionally, Cryptochromes 1 and 2 act cooperatively in repressing the transcription of PHYA and antagonistically on the transcription of PHYB2, which is promoted by CRY1a and repressed by CRY2 (Facella et al., 2008). In *Arabidopsis*, there is evidence for a direct interaction between PHYA and CRY1, with PHYA mediating a light-dependent phosphorylation of CRY1 (Ahmad et al., 1998), and between PHYB and CRY2, with the CRY2 probably suppressing PHYB signaling (Mas et al., 2000). In tomato, an additional level of suppression of PHYB signaling could be represented by the repressive action of CRY1a and CRY2 on PHYB2 transcript levels. Another effect is the approximately 3–10-fold increase of PHYF transcripts in plants lacking a functional CRY1a.

The fact that under constant light conditions all tomato cryptochromes plus PHYA, PHYB2, and PHYF seem to keep their oscillations following a period close to 24 hours, though with lower amplitude and minor changes in the phase of the mRNA abundance peaks (Facella et al., 2008), hints that a circadian clock regulates the expression of these photoreceptors, as seen in the *Arabidopsis* closest homologs, PHYA, PHYD, and PHYF (Tòth et al., 2001). In contrast, PHYB1 and PHYE lose their rhythmicity in light constant conditions, while the most closely related *Arabidopsis* homologs, PHYB and PHYE, continue to cycle in light constant conditions with a peak at the beginning or in the first half of the light phase (Tòth et al., 2001). The different regulation in the two plants could reflect the different functional organization of the photoreceptor gene families. Differently from *Arabidopsis*, tomato flowering is day-neutral. In addition, the evident arhythmicity of PHYA and PHYB2 caused by the over-expression of CRY2 (Facella et al., 2008) is quite intriguing and suggests that this condition specifically disrupts the output signal from the clock to PHYA and PHYB2.

In tomato, cryptochromes are also involved in hormone–photoreceptor crosstalk (Facella et al., 2012a, b). The molecular effects of exogenous GA_3 auxin (IAA) and ABA on CRYs and PHYs transcripts in WT tomato, as well as in mutant genotype *cry1a-* (Weller et al., 2001) and in transgenic line *CRY2OX* (Giliberto et al., 2005), were investigated in a recent study by Facella and collaborators (Facella et al., 2012a, b). The transcription pattern of the phytochrome gene family, following treatment with GA_3, evidenced an opposite response in *cry1a-* plants with respect to WT and *CRY2OX* tomatoes (Facella et al., 2012a, b). Indeed, when a functional form of CRY1a protein is absent, all five phytochromes are constantly down-regulated, while, when CRY1a works normally (in WT and *CRY2OX* plants), the same genes appear to be mostly up-regulated (Facella et al., 2012a, b). This evidence demonstrated that exogenous GA_3, in tomato, is able to modify the diurnal expression pattern of several photoreceptor genes, especially when a working form of cryptochrome 1a is absent. These results suggest the existence, in tomato, of a molecular network among cryptochrome 1a, GA_3, and the other photoreceptor genes. The photoreceptor response to IAA treatment is lower than for gibberellin (Facella et al., 2012a, b). Once again, the most sensitive genotype to exogenous hormone is clearly *cry1a-*, especially when focusing on the cryptochrome mRNA transcripts: CRY1a, CRY1b, and CRY2 are down-regulated. In *WT* and *CRY2OX* plants no clear pattern of up- or down-regulation of cryptochrome transcripts was observed. CRY1a may play a crucial role in the regulation of cryptochrome expression under auxin stimulus as well; however, this role seems to be absent for phytochromes, which are almost totally unaffected in *cry1a*- plants. Therefore, the action of CRY1a on tomato photoreceptor gene transcripts changes according to different hormonal stimuli (Facella et al., 2012a, b). Generally, ABA does not have dramatic effects on transcription of cryptochrome genes (Facella et al., 2012a, b).

6.5 References

Ahmad M, Cashmore AR. 1993. HY4 gene of A. *thaliana* encodes a protein with characteristics of a blue-light photoreceptor. *Nature* 366: 162–6.

Ahmad M, Jarillo JA, Smirnova O, Cashmore AR. 1998. The CRY1 blue light photoreceptor of *Arabidopsis* interacts with phytochrome A in vitro. *Mol Cell* 1: 939–48.

Blasing OE, Gibon Y, Gunther M, Hohne M, Morcuende R, et al. 2005. Sugars and circadian regulation make major contributions to the global regulation of diurnal gene expression in *Arabidopsis*. *Plant Cell* 17: 3257–81.

Canamero RC, Bakrim N, Bouly JP, Garay A, Dudkin EE, et al. 2006. Cryptochrome photoreceptors CRY1 and CRY2 antagonistically regulate primary root elongation in *Arabidopsis thaliana*. *Planta* 224: 995–1003.

Chaves I, Pokorny R, Byrdin M, Hoang N, Ritz T, et al. 2011. The cryptochromes: blue light photoreceptors in plants and animals. *Annu Rev Plant Biol* 62: 335–64.

Chen M, Chory J, Fankhauser C. 2004. Light signal transduction in higher plants. *Annu Rev Genet* 38: 87–117.

Colon-Carmona A, Chen DL, Yeh KC, Abel S. 2000. Aux/IAA proteins are phosphorylated by phytochrome in vitro. *Plant Physiol* 124: 1728–38.

Corbesier L, Coupland G. 2006. The quest for florigen: a review of recent progress. *J Exp Bot* 57: 3395–403.

Covington MF, Maloof JN, Straume M, Kay SA, Harmer SL. 2008. Global transcriptome analysis reveals circadian regulation of key pathways in plant growth and development. *Genome Biol* 9: R130.

Danon A, Coll NS, Apel K. 2006. Cryptochrome-1-dependent execution of programmed cell death induced by singlet oxygen in *Arabidopsis thaliana*. *Proc Natl Acad Sci USA* 103: 17036–41.

de Montaigu A, Tóth R, Coupland G. 2010. Plant development goes like clockwork. *Trends Genet* 26: 296–306.

Devlin PF, Kay SA. 2000. Cryptochromes are required for phytochrome signaling to the circadian clock but not for rhythmicity. *Plant Cell* 12: 2499–510.

Dunlap JC. 1999. Molecular bases for circadian clocks. *Cell* 96: 271–90.

Eberhard S, Finazzi G, Wollman FA. 2008. The dynamics of photosynthesis. *Annu Rev Genet* 42: 463–515.

El-Assal SE, Alonso-Blanco C, Hanhart CJ, Koornneef M. 2004. Pleiotropic effects of the *Arabidopsis* cryptochrome 2 allelic variation underlie fruit trait-related QTL. *Plant Biol* 6: 370–4.

Facella P, Lopez L, Chiappetta A, Bitonti MB, Giuliano G, et al. 2006. CRY-DASH gene expression is under the control of the circadian clock machinery in tomato. *FEBS Lett* 580: 4618–24.

Facella P, Lopez L, Carbone F, Galbraith DW, Giuliano G, et al. 2008. Diurnal and circadian rhythms in the tomato transcriptome and their modulation by cryptochrome photoreceptors. *PLoS One* 3: e2798.

Facella P, Daddiego L, Giuliano G, Perrotta G. 2012a. Gibberellin and auxin influence the diurnal transcription pattern of photoreceptor genes via CRY1a in tomato. *PLoS One* 7: 30121.

Facella P, Daddiego L, Perrotta G. 2012b. CRY1a influences the diurnal transcription of photoreceptor genes in tomato plants after gibberellin treatment. *Plant Signal Behav* 7: 1–3.

Fankhauser C, Chory J. 1997. Light control of plant development. *Annu Rev Cell Dev Biol* 13: 203–29.

Fankhauser C, Staiger D. 2002. Photoreceptors in *Arabidopsis thaliana*: light perception signal transduction and entrainment of the endogenous clock. *Planta* 216: 1–16.

Folta KM, Pontin MA, Karlin-Neumann G, Bottini R, Spalding EP. 2003. Genomic and physiological studies of early cryptochrome 1 action demonstrate roles for auxin and gibberellin in the control of hypocotyl growth by blue light. *Plant J* 36: 203–14.

Foo E, Ross JJ, Davies NW, Reid JB, Weller JL. 2006. A role for ethylene in the phytochrome-mediated control of vegetative development. *Plant J* 46: 911–21.

Franklin KA, Quail PH. 2010. Phytochrome functions in *Arabidopsis* development. *J Exp Bot* 61(1): 11–24.

Fuglevand G, Jackson JA, Jenkins GI. 1996. UV-B, UV-A, and blue light signal transduction pathways interact synergistically to regulate chalcone synthase gene expression in *Arabidopsis*. *Plant Cell* 8: 2347–57.

Giliberto L, Perrotta G, Pallara P, Weller JL, Fraser PD, et al. 2005. Manipulation of the blue light photoreceptor cryptochrome 2 in tomato affects vegetative development, flowering time, and fruit antioxidant content. *Plant Physiol* 137: 199–208.

Graf A, Schlereth A, Stitt M, Smith AM. 2010. Circadian control of carbohydrate availability for growth in *Arabidopsis* plants at night. *Proc Natl Acad Sci USA* 107: 9458–63.

Guo H, Yang H, Mockler TC, Lin C. 1998. Regulation of flowering time by *Arabidopsis* photoreceptors. *Science* 279: 1360–3.

Harmer SL. 2009. The circadian system in higher plants. *Annu Rev Plant Biol* 60: 357–77.

Harmer SL, Hogenesch JB, Straume M, Chang HS, Han B, et al. 2000. Orchestrated transcription of key pathways in *Arabidopsis* by the circadian clock. *Science* 290, 2110–13.

Hauser BA, CordonnierPratt MM, DanielVedele F, Pratt LH. 1995. The phytochrome gene family in tomato includes a novel subfamily. *Plant Mol Biol* 29: 1143–55.

Hayama R, Coupland G. 2004. The molecular basis of diversity in the photoperiodic flowering responses of *Arabidopsis* and rice. *Plant Physiol* 135: 677–84.

Hotta CT, Gardner MJ, Hubbard KE, Baek SJ, Dalchau N, et al. 2007. Modulation of environmental responses of plants by circadian clocks. *Plant Cell Environ* 30: 333–49.

Huang Y, Baxter R, Smith BS, Partch CL, Colbert CL, et al. 2006. Crystal structure of cryptochrome 3 from *Arabidopsis thaliana* and its implications for photolyase activity. *Proc Natl Acad Sci USA* 103: 17701–6.

Ishikawa M, Kiba T, Chua NH. 2006. The Arabidopsis SPA1 gene is required for circadian clock function and photoperiodic flowering. *Plant J* 46(5): 736–46.

Jang S, Marchal V, Panigrahi KC, Wenkel S, Soppe W, et al. 2008. *Arabidopsis* COP1 shapes the temporal pattern of CO accumulation conferring a photoperiodic flowering response. *EMBO J* 27: 1277–88.

Jenkins GI. 2009. Signal transduction in responses to UV-B radiation. *Annu Rev Plant Biol* 60: 407–31.

Jensen PJ, Hangarter RP, Estelle M. 1998. Auxin transport is required for hypocotyl elongation in light-grown but not dark-grown *Arabidopsis*. *Plant Physiol* 116: 455–62.

Jeong RD, Chandra-Shekara AC, Barman SR, Navarre D, Klessig DF, et al. 2010. Cryptochrome 2 and phototropin 2 regulate resistance protein-mediated viral defense by negatively regulating an E3 ubiquitin ligase. *Proc Natl Acad Sci USA* 107: 13538–43.

Jiao Y, Ma L, Strickland E, Deng XW. 2005. Conservation and divergence of light-regulated genome expression patterns during seedling development in rice and *Arabidopsis*. *Plant Cell* 17: 3239–56.

Jiao Y, Lau OS, Deng XW. 2007. Light-regulated transcriptional networks in higher plants. *Nat Rev Genet* 8: 217–030.

Kang CY, Lian HL, Wang FF, Huang JR, Yang HQ. 2009. Cryptochromes, phytochromes, and COP1 regulate light-controlled stomatal development in *Arabidopsis*. *Plant Cell* 21: 2624–41.

Kasemir H. 1979. Control of chloroplast formation by light. *Cell Biol Int Rep* 3(3): 197–214.

Keller MM, Jaillais Y, Pedmale UV, Moreno JE, Chory J, et al. 2011. Cryptochrome 1 and phytochrome B control shade-avoidance responses in *Arabidopsis* via partially independent hormonal cascades. *Plant J* 67: 195–207.

Kerckhoffs LHJ, Kelmenson PM, Schreuder MEL, Kendrick CI, Kendrick RE, et al. 1999. Characterization of the gene encoding the apoprotein of phytochrome B2 in tomato, and identification of molecular lesions in two mutant alleles. *Mol Gen Genet* 261: 901–7.

Kleine T, Kleine T, Lockhart P, Batschauer A. 2003. An *Arabidopsis* protein closely related to Synechocystis cryptochrome is targeted to organelles. *Plant J* 35: 93–103.

Kleine T, Kindgren P, Benedict C, Hendrickson L, Strand A. 2007. Genome-wide gene expression analysis reveals a critical role for CRYPTOCHROME1 in the response of *Arabidopsis* to high irradiance. *Plant Physiol* 144: 1391–406.

Kuno N, Furuya M. 2000. Phytochrome regulation of nuclear gene expression in plants. *Semin Cell Dev Biol* 11: 485–93.

Lee J, He K, Stolc V, Lee H, Figueroa P, et al. 2007. Analysis of transcription factor HY5 genomic binding sites revealed its hierarchical role in light regulation of development. *Plant Cell* 19: 731–49.

Li J, Li G, Wang H, Deng XW. 2011. Phytochrome signaling mechanisms. *Arabidopsis* Book 9, e0148.

Lian HL, He SB, Zhang YC, Zhu DM, Zhang JY, et al. 2011. Blue-light-dependent interaction of cryptochrome 1 with SPA1 defines a dynamic signaling mechanism. *Genes Dev* 25: 1023–8.

Lidder P, Gutierrez RA, Salome PA, McClung CR, Green PJ. 2005. Circadian control of messenger RNA stability: association with a sequence-specific messenger RNA decay pathway. *Plant Physiol* 138: 2374–85.

Lin C, Shalitin D. 2003. Cryptochrome structure and signal transduction. *Annu Rev Plant Biol* 54: 469–96.

Linka M, Weber AP. 2005. Shuffling ammonia between mitochondria and plastids during photorespiration. *Trends Plant Sci* 10: 461–5.

Liu B, Zuo Z, Liu H, Liu X, Lin C. 2011. *Arabidopsis* cryptochrome 1 interacts with SPA1 to suppress COP1 activity in response to blue light. *Genes Dev* 25: 1029–34.

Liu H, Yu X, Li K, Klejnot J, Yang H, et al. 2008. Photoexcited CRY2 interacts with CIB1 to regulate transcription and floral initiation in *Arabidopsis*. *Science* 322: 1535–9.

Liu H, Liu B, Zhao C, Pepper M, Lin C. 2011. The action mechanisms of plant cryptochromes. *Trends Plant Sci* 16: 684–91.

Locke JC, Kozma-Bognar L, Gould PD, Feher B, Kevei E, et al. 2006. Experimental validation of a predicted feedback loop in the multi-oscillator clock of *Arabidopsis thaliana*. *Mol Syst Biol* 2: 1–6.

Lopez L, Carbone F, Bianco L, Giuliano G, Facella P, et al. 2012. Tomato plants overexpressing cryptochrome 2 reveal altered expression of energy and stress-related gene products in response to diurnal cues. *Plant Cell Environ* 35: 994–1012.

Lopez-Juez E, Dillon E, Magyar Z, Khan S, Hazeldine S, et al. 2008. Distinct light-initiated gene expression and cell cycle programs in the shoot apex and cotyledons of *Arabidopsis*. *Plant Cell* 20: 947–68.

Ma L, Li J, Qu L, Hager J, Chen Z, et al. 2001. Light control of *Arabidopsis* development entails coordinated regulation of genome expression and cellular pathways. *Plant Cell* 13: 2589–607.

Mao J, Zhang YC, Sang Y, Li QH, Yang HQ. 2005. A role for *Arabidopsis* cryptochromes and COP1 in the regulation of stomatal opening. *Proc Natl Acad Sci USA* 102: 12270–5.

Mas P, Devlin PF, Panda S, Kay SA. 2000. Functional interaction of phytochrome B and cryptochrome 2. *Nature* 408: 207–11.

McClung CR. 2010. A modern circadian clock in the common angiosperm ancestor of monocots and eudicots. *BMC Biol* 8: 55.

McWatters HG, Devlin PF. 2011. Timing in plants--a rhythmic arrangement. *FEBS Lett* 585: 1474–84.

Michael TP, Mockler TC, Breton G, McEntee C, Byer A, et al. 2008. Network discovery pipeline elucidates conserved time-of-day-specific cis-regulatory modules. *PLoS Genet* 4: e14.

Mochizuki T, Onda Y, Fujiwara E, Wada M, Toyoshima Y. 2004. Two independent light signals cooperate in the activation of the plastid psbD blue light-responsive promoter in *Arabidopsis*. *FEBS Lett* 571: 26–30.

Möglich A, Yang X, Ayers RA, Moffat K. 2010. Structure and function of plant photoreceptors. *Annu Rev Plant Biol* 2010; 61: 21–47.

Nagashima A, Hanaoka M, Shikanai T, Fujiwara M, Kanamaru K, et al. 2004. The multiple-stress responsive plastid sigma factor, SIG5, directs activation of the psbD blue light-responsive promoter (BLRP) in *Arabidopsis thaliana*. *Plant Cell Physiol* 45: 357–68.

Neff MM, Street IH, Turk EM, Ward JM. 2006. Interaction of light and hormone signalling to mediate photomorphogenesis. In: Schafer E, Nagy F (Eds) *Photomorphogenesis in Plants and Bacteria*. Springer, Dordrecht, the Netherlands. pp. 439–73.

Nelson DC, Lasswell J, Rogg LE, Cohen MA, Bartel B. 2000. FKF1, a clock controlled gene that regulates the transition to flowering in *Arabidopsis*. *Cell* 101: 331–40.

Ninu L, Ahmad M, Miarelli C, Cashmore AR, Giuliano G. 1999. Cryptochrome 1 controls tomato development in response to blue light. *Plant J* 18: 551–6.

Ohgishi M, Saji K, Okada K, Sakai T. 2004. Functional analysis of each blue light receptor, cry1, cry2, phot1, and phot2, by using combinatorial multiple mutants in *Arabidopsis*. *Proc Natl Acad Sci USA* 101: 2223–8.

Ozsolak F, Milos PM. 2011. RNA sequencing: advances, challenges and opportunities. *Nat Rev Genet* 12: 87–98.

Partch CL, Clarkson MW, Ozgür S, Lee AL, Sancar A. 2005. Role of structural plasticity in signal transduction by the cryptochrome blue-light photoreceptor. *Biochemistry* 44: 3795–805.

Pedmale UV, Celaya RB, and Liscum E. 2010. Phototropism: mechanism and outcomes. *Arabidopsis Book* 8, e0125.

Perrotta G, Ninu L, Flamma F, Weller JL, Kendrick RE, et al. 2000. Tomato contains homologues of *Arabidopsis* cryptochrome 1 and 2. *Plant Mol Biol* 42: 765–73.

Perrotta G, Yahoubian G, Nebuloso E, Renzi L, Giuliano G. 2001. Tomato and barley contain duplicated copies of cryptochrome 1. *Plant Cell Environ* 24, 991–7.

Peschke F, Kretsch T. 2011. Genome-wide analysis of light-dependent transcript accumulation patterns during early stages of *Arabidopsis* seedling deetiolation. *Plant Physiol* 155(3): 1353–66.

Pokorny R, Klar T, Hennecke U, Carell T, Batschauer A, et al. 2008. Recognition and repair of UV lesions in loop structures of duplex DNA by DASH-type cryptochrome. *Proc Natl Acad Sci USA* 105: 21 023–7.

Pratt LH, Cordonnier-Pratt MM, Hauser B, Caboche M. 1995. Tomato contains two differentially expressed genes encoding B-type phytochromes, neither of which can be considered an ortholog of *Arabidopsis* phytochrome B. *Planta* 197: 203–6.

Pruneda-Paz JL, Breton G, Para A, Kay SA. 2009. A functional genomics approach reveals CHE as a component of the *Arabidopsis* circadian clock. *Science* 323: 1481–5.

Rutitzky M, Ghiglione HO, Curá JA, Casal JJ, Yanovsky MJ. 2009. Comparative genomic analysis of light-regulated transcripts in the Solanaceae. *BMC Genomics* 10: 60.

Schaffer R, Landgraf J, Accerbi M, Simon V, Larson M, et al. 2001. Microarray analysis of diurnal and circadian-regulated genes in *Arabidopsis*. *Plant Cell* 13: 113–23.

Searle I, Coupland G. 2004. Induction of flowering by seasonal changes in photoperiod. *EMBO J* 23: 1217–22.

Sellaro R, Hoecker U, Yanovsky M, Chory J, Casal JJ. 2009. Synergism of red and blue light in the control of *Arabidopsis* gene expression and development. *Curr Biol* 19: 1216–20.

Somers DE, Devlin PF, Kay SA. 1998. Phytochromes and cryptochromes in the entrainment of the *Arabidopsis* circadian clock. *Science* 282: 1488–90.

Symons GM, Reid JB. 2003. Hormone levels and response during de-etiolation in pea. *Planta* 216: 422–31.

Tepperman JM, Zhu T, Chang H-S, Wang X, Quail PH. 2001. Multiple transcription-factor genes are early targets of phytochrome A signaling. *Proc Natl Acad Sci USA* 98: 9437–42.

Tepperman JM, Hwang YS, Quail PH. 2006. phyA dominates in transduction of red-light signals to rapidly responding genes at the initiation of *Arabidopsis* seedling de-etiolation. *Plant J* 48: 728–42.

Thum KE, Kim M, Christopher DA, Mullet JE. 2001. Cryptochrome 1, cryptochrome 2, and phytochrome A co-activate the chloroplast psbD blue light-responsive promoter. *Plant Cell* 13: 2747–60.

Tomato Genome Consortium. 2012. The tomato genome sequence provides insights into fleshy fruit evolution. *Nature* 485: 635–41.

Tòth R, Kevei E, Hall A, Millar AJ, Nagy F, et al. 2001. Circadian clock-regulated expression of phytochrome and cryptochrome genes in *Arabidopsis*. *Plant Physiol* 127: 1607–16.

Tsuchida-Mayama T, Sakai T, Hanada A, Uehara Y, Asami T, et al. 2010. Role of the phytochrome and cryptochrome signaling pathways in hypocotyl phototropism. *Plant J* 62: 653–62.

Tsunoyama Y, Morikawa K, Shiina T, Toyoshima Y. 2002. Blue light specific and differential expression of a plastid sigma factor, Sig5 in *Arabidopsis thaliana*. *FEBS Lett* 516: 225–8.

Ulm R, Baumann A, Oravecz A, Máté Z, Adám E, et al. 2004. Genome-wide analysis of gene expression reveals function of the bZIP transcription factor HY5 in the UV-B response of *Arabidopsis*. *Proc Natl Acad Sci USA* 101: 1397–402.

Valverde F, Mouradov A, Soppe W, Ravenscroft D, Samach A, et al. 2004. Photoreceptor regulation of CONSTANS protein in photoperiodic flowering. *Science* 303: 1003–6.

Vandenbussche F, Habricot Y, Condiff AS, Maldiney R, Van der Straeten D, et al. 2007. HY5 is a point of convergence between cryptochrome and cytokinin signalling pathways in *Arabidopsis thaliana*. *Plant J* 49: 428–41.

Vandenbussche F, Pierik R, Millenaar FF, Voesenek LA, Van Der Straeten D. 2005. Reaching out of the shade. *Curr Opin Plant Biol* 8: 462–8.

Van Tuinen A, Kerckhoffs LH, Nagatani A, Kendrick RE, Koornneef M. 1995a. Far-red light-insensitive, phytochrome A-deficient mutants of tomato. *Mol Gen Genet* 246: 133–41.

Van Tuinen A, Kerckhoffs L, Nagatani A, Kendrick RE, Koornneef M. 1995b. A Temporarily red light-insensitive mutant of tomato lacks a light-stable, B-like phytochrome. *Plant Physiol* 108: 939–47.

Wang H, Ma LG, Li JM, Zhao HY, Deng XW. 2001. Direct interaction of *Arabidopsis* cryptochromes with COP1 in light control development. *Science* 294: 154–8.

Weller JL, Schreuder MEL, Smith H, Koornneef M, Kendrick RE. 2000. Physiological interactions of phytochromes A, B1 and B2 in the control of development in tomato. *Plant J* 24: 345–56.

Weller JL, Perrotta G, Schreuder ME, van Tuinen A, Koornneef M, et al. 2001. Genetic dissection of blue-light sensing in tomato using mutants deficient in cryptochrome 1 and phytochromes A, B1 and B2. *Plant J* 25: 427–40.

Wu G, Spalding EP. 2007. Separate functions for nuclear and cytoplasmic cryptochrome 1 during photomorphogenesis of *Arabidopsis* seedlings. *Proc Natl Acad Sci USA* 104, 18813–18.

Wu L, Yang H. 2010. CRYPTOCHROME 1 is implicated in promoting R protein-mediated plant resistance to Pseudomonas syringae in *Arabidopsis*. *Mol Plant* 3: 539–48.

Xu P, Xiang Y, Zhu H, Xu H, Zhang Z, et al. 2009. Wheat cryptochromes: subcellular localization and involvement in photomorphogenesis and osmotic stress responses. *Plant Physiol* 149: 760–74.

Yamaguchi S. 2008. Gibberellin metabolism and its regulation. *Annu Rev Plant Biol* 59: 225–51.

Yanovsky MJ., Mazzella MA, Whitelam GC, Casal JJ. 2001. Resetting of the circadian clock by phytochromes and cryptochromes in *Arabidopsis*. *J Biol Rhythms* 16: 523–30.

Yu X, Klejnot J, Zhao X, Shalitin D, Maymon M, et al. 2007. *Arabidopsis* cryptochrome 2 completes its posttranslational life cycle in the nucleus. *Plant Cell* 19, 3146–56.

Yu X, Liu H, Klejnot J, Lin C. 2010. The cryptochrome blue light receptors. *Arabidopsis* Book 8, e0135.

Zeng J, Wang Q, Lin J, Deng K, Zhao X, et al. 2010. *Arabidopsis* cryptochrome-1 restrains lateral roots growth by inhibiting auxin transport. *J Plant Physiol* 167: 670–3.

Zhang H, He H, Wang X, Wang X, Yang X, et al. 2011. Genome-wide mapping of the HY5-mediated gene networks in *Arabidopsis* that involve both transcriptional and post-transcriptional regulation. *Plant J* 65: 346–58.

Zhao X, Yu X, Foo E, Symons GM, Lopez J, et al. 2007. A study of gibberellin homeostasis and cryptochrome-mediated blue light inhibition of hypocotyl elongation. *Plant Physiol* 145: 106–18.

Zhu Z, Guo H. 2008. Genetic basis of ethylene perception and signal transduction in *Arabidopsis. J Integr Plant Biol* 50: 808–15.

Zuo Z, Liu H, Liu B, Liu X, Lin C. 2011. Blue light-dependent interaction of CRY2 with SPA1 regulates COP1 activity and floral initiation in *Arabidopsis. Curr Biol* 21: 841–7.

7

Genomics of grapevine: from genomics research on model plants to crops and from science to grapevine breeding

Fatemeh Maghuly, BOKU VIBT, Austria, Giorgio Gambino, Plant Virology Institute, National Research Council (IVV-CNR), Italy, Tamás Deák, Corvinus University of Budapest, Hungary, and Margit Laimer, BOKU VIBT, Austria

DOI: 10.1533/9781908818478.119

Abstract: The grape breeding industry is dynamic, with new cultivars being released annually. In recent years there has been interest in breeding new grape cultivars to develop a new gene with wider ecological adaptation and to provide a response to viral pathogens. To achieve these goals, biotechnological approaches to plant disease are employed. Genomic studies and genetic markers can significantly speed up the selection of resistant seedlings. Natural virus resistance may be exploited by transfer of genes from resistant genotypes. Plant transformation has become an essential tool for plant molecular biologists, and genetically improved plants are the focus of many breeding programs. Integration of genes from diverse biological sources into grape genomes promises to broaden the gene pool and tailor plant varieties for specific traits. This chapter presents the molecular characterization and localization of T-DNA insertion in grapevines carrying genes of viral origin. Detailed analyses have revealed the crucial factors that influence transgene expression. Plants appear to respond to infection by activating a post-transcriptional gene silencing mechanism.

Key words: grapevine, cultivars, genomics, breeding, biotechnology, viruses, resistance, molecular markers, transgene silencing, insertion loci, integration of vector backbone, stability of inserted transgenes, DNA methylation.

7.1 Use of genetic and molecular markers for studies of genetic diversity and genome selection in grapevine

Genetically, domestication of fruit trees involves a change in the reproductive biology of the plants by shifting from sexual reproduction to vegetative propagation (Zohary and Spiegel-Roy, 1975). Cultivated varieties of fruit trees are maintained vegetatively by cuttings, rooting of twigs or suckers. These methods are in sharp contrast with the life cycle of their wild relatives (with high levels of heterozygosity), which reproduce from seeds. Man-made selection has led to some improved cultivars with enhancement of some agronomic traits, but at the cost of reduction of genetic variability (Maghuly et al., 2005; Zohary and Hopf, 1993), which may result in inbreeding depression. To achieve certain breeding goals, particularly for a wider ecological adaptation, disease resistance and novel fruit quality traits, the use of germplasm from different groups and eco-geographical regions is necessary. Therefore the identification of individual accessions, among a representative set of commercially available varieties, represents a prerequisite for setting any breeding strategy for fruit trees such as grape (Maghuly et al. 2005, 2006a, b).

Grapevine was domesticated 6000–8000 years ago (McGovern, 2003), representing a remarkable history of cultivation. Today about 6000–10 000 vegetatively propagated cultivars are assumed to exist worldwide in germplasm collections (Galet, 2000; Laucou et al., 2011). The genetic variability of the European grapevine, *Vitis vinifera*, is comparable with highly variable plant species such as maize (Lijavetzky et al., 2007; Myles et al., 2011). During the past 100 years, the diversity of cultivated grapevines has been boosted with the introduction of North American *Vitis* species into cultivation and breeding as sources of resistance against different pests (e.g. phylloxera), diseases (e.g. mildews) and environmental stresses (e.g. winter hardiness) (Reisch et al., 2012). Therefore, assessment and mapping of genetic variability of cultivated and native grape species are of primary importance for marker-assisted breeding (This et al., 2006).

On the other hand, a large number of fruit tree cultivars are commercialized, and the breeding industry is particularly prompt in the release of new cultivars. This leads to the existence of a high number of cultivars of economic value with different synonyms whose distinction demands the use of fast and reliable techniques for molecular fingerprinting (Bassi et al., 1995; Egea et al., 1999; Maghuly et al., 2006b, c, 2007). Molecular markers are valuable tools for the genotyping of accessions, for the classification and management of grapevine germplasm collections, and for identifying potentially promising parental genotypes for breeding purposes. Furthermore, in a breeding program the correct and reliable identification of accessions is highly important. This allows the design of crosses that maximize genetic variability with the objective of obtaining new genotypes, for example, resistant to viruses, that could be cultivated in a wide geographical area.

Although recent grape breeding programs may differ in methodology (clonal selection, cross-breeding, biotechnology), aims (abiotic or biotic resistance, quality) or even subjects (wine grapes, table grapes, rootstocks), identification of genotypes used during the process is of primary importance in all cases. Crosses can then be carefully designed according to the breeding goals. In several cases haplotypes of genes or quantitative trait loci (QTLs) instead of varieties or hybrids are selected for parental genotypes during the planning period, when the utilization of molecular markers is of significant assistance for the breeder.

A breakthrough in DNA-based cultivar identification in grapevines started after the first studies of simple sequence repeat (SSR) or microsatellite loci for grape varieties (Thomas et al., 1993). It has been shown that these markers are inherited in a co-dominant Mendelian manner, making them highly applicable for genetic mapping and for the study of genetic relatedness (Thomas et al., 1994). After early works on common varieties, several countries and research groups started to develop new microsatellite markers (e.g. Adam-Blondon et al., 2004; Arroyo-Garcia and Martinez-Zapater, 2004; Bowers et al., 1996; Di Gaspero et al., 2005; Goto-Yamamoto et al., 2006; Merdinoglu et al., 2005; Sefc et al., 1999) and to prepare SSR databanks of local and international varieties. The efforts were joined by setting up the framework of the *Vitis* Microsatellite Consortium (VMC, 1997–2004). As a result of these efforts, a set of six standard microsatellite loci was proposed for the identification and passport description of grape varieties (This et al., 2004).

After initial screening of germplasm collections it quickly became clear that many synonyms or homonyms exist among genotypes (This et al., 2006). This was followed by extensive surveys of national collections, in which thousands of accessions were profiled based on several SSR loci (Cipriani et al., 2010; Laucou et al., 2011). These studies clarified ambiguities in naming and registering grape genotypes. As this work was carried out on thousands of genotypes, detailed pedigree information for many varieties also became available, suggesting parent–offspring relationships of newly bred and also of traditional cultivars.

During recent decades, several molecular markers linked to agronomically important traits have been identified using traditional linkage mapping or by analyzing the genetic background of mutant phenotypes. Genetic determinants were identified for berry color (Doligez et al., 2002; Kobayashi et al., 2004), flesh development (Fernandez et al., 2006) and cluster structure (Fernandez et al., 2010). Major quantitative traits were mapped and tightly linked molecular markers were identified for seedlessness and berry weight (Cabezas et al., 2006; Costantini et al., 2008; Doligez et al., 2002), berry flavors and odor components (Battilana et al., 2009; Doligez et al., 2006; Duchêne et al., 2009), and disease resistance against powdery mildew (Dalbó et al., 2001; Hoffmann et al., 2008; Mahanil et al., 2012; Welter et al., 2007), Pierce's disease (Riaz et al., 2008), crown gall disease (Kuczmog et al., 2012) and the dagger nematode *Xiphenema index* (Xu et al., 2008). Although these traits and markers can be used in marker-assisted selection (MAS), the full potential of marker-assisted selection (MAS) will remain untapped until more complex traits become available to predict.

The almost simultaneous publication of a homozygous (Jaillon et al., 2007) and a heterozygous (Velasco et al., 2007) genome of 'Pinot noir' and the availability of dense physical maps (Moroldo et al., 2008; Troggio et al., 2007) opened new possibilities both for the development of molecular markers and for the assessment of genetic diversity. After resequencing of several grape genotypes, single nucleotide polymorphism (SNP) arrays were designed with 9000 valid polymorphisms (Myles et al., 2010, 2011). Such high throughput methods as SNP arrays and the still rapidly evolving next-generation sequencing (NGS) technologies (reviewed by Hamilton and Buell, 2012) allow not only to gain new insights into grape regulatory pathways (Pantaleo et al., 2010) but also to get an in-depth understanding of genetic factors defining complex traits. Genome wide association mapping (McCarthy et al., 2008) and genomic selection (Heffner et al., 2009) are promising techniques to aid marker-assisted breeding.

7.2 Grapevine breeding

Plant breeding is defined as identifying and selecting desirable traits and combining these into individual plants. Since their rediscovery in 1900 by De Vries, Correns and Tschermak, the Mendelian rules (Mendel, 1866) of genetics provide the scientific basis for plant breeding. Research based on these rules of inheritance revealed that the phenotypic traits of plants are controlled by genes located on chromosomes. Therefore, plant breeding can be considered as the manipulation of genomes by producing new combinations of genes and chromosomes. The two major approaches for genetic improvement are conventional breeding and genetic engineering.

In conventional breeding the pools of available genes, and the traits they code for, are limited due to sexual incompatibility with other related crop species and with their wild relatives. By conventional breeding, natural virus resistance genes may be exploited by the identification of resistant genotypes in fruit trees (Bassi, 2006; Hartman and Neumüller, 2006), followed by the transfer of resistance genes into new germplasm through hybridization. Genomic studies and genetic markers can speed up the selection of putative resistant seedlings (Decroocq et al., 2005, 2006; Dondini et al., 2007). However, the process from hybridization to cultivar release can span decades.

In grapevine, the most extensive breeding programs aim at the incorporation of multiple fungal resistance into the genetic background of high-quality European grape varieties. As the pooling of multiple resistance sources against one pathogen is desired, application of MAS is inevitable (Eibach et al., 2007). However, this genetic approach must rely on high-throughput phenotyping (Finkel, 2009; Tester and Langridge, 2010). Current breeding programs often involve repeated backcrosses after the introgression of a desired trait (e.g. disease resistance) of wild grape species into the genetic background of high-quality varieties. Although MAS for the gene of interest is possible in many cases, the remaining part of the genome from the donor parent should be eliminated as quickly and as precisely as possible. Identification of seedlings showing the desired recombination events around the gene(s) of interest and the highest proportion of recurrent genome background at the same time can be achieved by high-throughput, high-density genotyping of large backcross populations (Di Gaspero and Cattonaro, 2010). The next step for grape breeding would be the transfer of the existing theoretical background of such complex marker-assisted breeding programs into breeding practice.

On the other hand, conventional breeding for resistance to plant pathogens, for example, viruses, and/or to their vectors has long been used, scoring remarkable results only with a limited number of crops and viruses. Fewer than 180 genes have been utilized to date for introgressing monogenic or polygenic resistance to crop plants, while there have been about a dozen cases of effective durable resistance (lasting up to 50 years), involving no more than ten viruses (Fraser, 1990; Khetarpal et al., 1998). The breeding and cultivation of virus-resistant plants are a major contribution to the control of viral diseases, since no chemical control strategies exist so far. Even if virus-free planting material is used for new plantation, the danger of contamination by neighboring plants is so high that currently only virus-resistant plants are considered a valid alternative (Hartmann, 1998; Ilardi and Nicola-Negri, 2011; Laimer et al., 2005). The major hurdles to conventional breeding for resistance are genetic incompatibility barriers, long breeding cycle and unavailability of natural resistance genes.

However, these problems can now be overcome by genetic improvement technology. The resistance obtained, commonly called 'non-conventional' or 'genetically improved', may be of the following types: (i) pathogen-derived resistance, conferred to a plant by genes isolated from the pathogen's genome, cloned, and engineered into the plant's genome (Laimer, 2003; Sanford and Johnston, 1985; Wally and Punja, 2010); (ii) resistance induced by other, non-microbial DNA sequences as plant-derived genes in the form of cisgenesis or intra-vector approach after identifying resistant cultivars (Bassi, 2006; Bhatti and Jha, 2010; Decroocq et al., 2005; Hartmann and Neumüller, 2006; Jacobsen and Schouten, 2007) and isolating resistance genes; or (iii) non-plant, non-pathogen transgenes, for example, plant antibodies and antiviral proteins (Prins et al., 2008; Safarnejad et al., 2012).

Work on pathogen-mediated resistance in woody species, focusing on virus resistance breeding in fruit trees and grapevines, started in 1988 (Laimer et al., 1991, 2009). No control of these pathogens by chemical means exists (CABI/EPPO 1991/1992), and the chemical control of their vectors, for example, aphids and nematodes, is ecologically questionable. The development of methods for producing genetically improved plants led virologists to devise novel virus resistance genes and use them to produce resistant plants (Laimer, 2007). One strategy was to simulate cross-protection by transforming plants with part of the viral genome, perhaps a single viral gene. But the question arose: which gene would be the strongest candidate? (Prins et al., 2008)

Powell et al. (1986) first used the coat protein gene of *Tobacco mosaic virus* (TMV) and discovered that the genetically improved plants had a greatly enhanced resistance to infection. Resistance was related to the level of expression of the viral coat protein and could be overcome by high doses of inoculum. Moreover, the plants had only a slightly enhanced resistance to infection by inoculation with TMV RNA (Nelson et al., 1987), suggesting that resistance resulted from inhibition of the presumed first step in virus replication, the dissociation of inoculum virus particles into protein and genomic RNA. The resistance was *tobamovirus*-specific; it was expressed strongly to *tobamoviruses* closely related to TMV (Nejidat and Beachy, 1990). This type of resistance was named coat protein-mediated resistance (CP-MR), the first example of pathogen-derived resistance (Sanford and Johnston, 1985) in plants. Transformation with coat protein genes also conferred resistance on a wide range of other plant viruses with RNA genomes, and the resistance was effective and durable under field conditions. When plants were transformed with untranslatable viral nucleotide sequences, resistance depended on gene silencing (Harrison and Robinson, 2005). Resistance to infection is expressed most strongly when the virus-derived transgene sequence is more than 1 kb (Longstaff et al., 1993) or the transgene transcript can form a hairpin structure containing a substantial base-paired stem (Waterhouse et al., 1998). Under such circumstances, resistance of genetically improved crops in the field can be strong, even where virus vectors are abundant (Fuchs and Gonsalves, 1995; Tricoli et al., 1995). Engineering virus resistance by genes encoding viral RNA-dependent RNA polymerase (RdRp) was also exploited by Golemboski et al. (1990) and Tenllado et al. (1996), using a truncated construct (30% of 54KDa RdRp) of the tobamovirus *Pepper mild mottle virus*, suggesting the coexistence of dual protein and RNA-mediated protection.

The successful use of pathogen derived resistance (PDR) to confer protection against plant viruses based on either protein- or RNA-mediated resistance led several research groups to investigate these approaches for resistance to *Grapevine fanleaf virus* (GFLV) and other viruses affecting grapevines. *Nepoviruses* are very important in the wine industry, causing significant reduction in grape yield and quality. As the virus is transferred by nematodes, it was assumed that developing resistant rootstocks would provide adequate protection against the virus. However, due to the lack of resistant germplasm in breeding programs, the genetically improved lines could be suitable to produce resistant planting material (Laimer et al., 2009). To achieve resistance against fanleaf disease, the CP genes of Arabis mosaic virus (ArMV) and grapevine fanleaf virus (GFLV) were used (Gölles et al., 2000; Vigne et al., 2004). Consideration regarding the safety of deliberate release of such genetically improved plants led to the development of different constructs, in which either no or a modified protein is produced; for example, in constructs with a GFLV-CP gene truncated at the 5'- or 3'-end, a non-translatable CP gene in sense or antisense orientation (Gölles et al., 2000). This construct should reduce the possible occurrence of heteroencapsidation (Robinson, 1996; Tepfer, 1993) or the formation of empty pseudovirions, described by Bertioli et al. (1991). Recombinant viruses were detected only when the transgene had no deletion in the 3'NTR adjacent to CP, concluding that a possible reduced genetically improved target length was involved in the prevention of recombination (Greene and Allison, 1994). It was suggested by Allison et al. (1996) that the replication initiation site should be excluded from the transgene construct. Untranslatable viral genes in the form of antisense constructs would assure that no additional protein is produced in the plants to minimize even putative biological risks of virus-resistant genetically modified plants. These approaches were first tested in herbaceous hosts and finally transferred to grapevines (Gölles et al., 2000). A delay in viral infection in transgenic grapevines in field trials was reported (Vigne et al., 2004). A decrease in virus titer was also shown in protoplasts of transgenic grapevine rootstock expressing the CP or movement protein (MP) gene of GFLV, assayed by protoplast electroporation (Valat et al., 2006). The artificial microRNA (amiRNA) based strategy was reported in grapevine (Jelly et al., 2012). They developed two amiRNA targeting the CP gene of GFLV.

Molecular characterization of 127 transgenic grapevine lines, later grouped into 39 independent transformation events, generated with six different transferred DNA (T-DNA) constructs of GFLV (Figure 7.1), confirmed the transgenity of all lines by PCR, Southern blot, enzyme-linked immunosorbent assay (ELISA) and real time-polymerase chain reaction (RT-PCR) (Maghuly et al., 2006c).

The fact that 46% of the plants carry a single copy insertion is important for further breeding purposes. Some of the lines also contain six copies of the transgene, which could be interesting for an experimental field trial, since more interference at the expression level might be responsible for the observed reduction in expression.

Further, accumulation of GFLV-CP could not be detected by ELISA in any of the grapevine lines tested, which is in accordance with the results reported by Gambino et al. (2005) and Spielmann et al. (2000). Indeed,

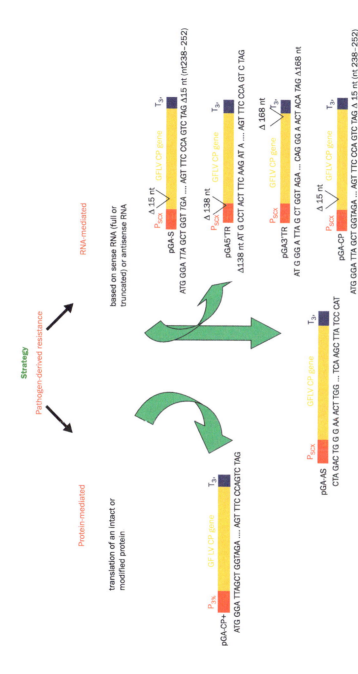

Figure 7.1 Expression cassettes of plant transformation vectors carrying different sequences of the GFLV-CP gene. Plasmid pGA-CP+ carries the full-length CP gene with an introduced start codon. pGA-CP with a deletion of 15 bp within the CP gene. pGA-AS contains the same gene construct as pGA-CP+ but in AS orientation. In the plasmid pGA-AS two stop codons were introduced after the inserted start codon. pGA-5'TR carries a CP which is shortened by 138 bp at the 5'-end. Plasmid pGA-3'TR contains a CP gene with a truncation of 168 bp at the 3'-end of the gene.

for a protection phenotype protein expression is not always required (Baulcombe, 2004).

From a safety point of view it might be advantageous to know that the expression of the CP construct is below the level of detection, thus minimizing the chance of heteroencapsidation, as described by Bertioli et al. (1991). The result of the RT-PCR findings clearly demonstrated that the transgene is fully transcribed (except in two lines) and no apparent correlation between T-DNA copy number and mRNA accumulation level exists.

Genetic engineering in grapevine, as in other plant species, does not always result in efficient transgene expression levels. Several cases have been reported to date in which the transgene copy number did not correlate with the level of transgene expression (Gelvin, 2003; Maghuly et al., 2006c, 2007). Two types of events can contribute to the occurrence of transgene silencing. The first is linked to the specific genomic site in which the T-DNA integrates itself. Indeed, integration into heterochromatic regions or into highly repetitive and methylated DNA sequences is correlated with reduced expression of the transgenes. On the other hand, the second type of event closely depends on the configuration of the integrated T-DNAs: multiple T-DNAs can integrate at one locus linked to each other in complex structures, which frequently results in transgene silencing (Gelvin, 2003; Ooms et al., 1982). Deeper investigation into T-DNA integration patterns in transgenic grapevines containing GFLV-CP genes (Gambino et al., 2009; Maghuly et al., 2006c) has provided important insights both on transgenic silencing in grapevine (Gambino et al., 2010) and on the integration process of foreign sequences within the grapevine genome (Gambino et al., 2009).

7.3 Transgene silencing

The process of gene silencing was first discovered in plants during recent years in unraveling the mechanism of RNA silencing, a process leading to the degradation of homologous mRNAs, also known as post-transcriptional gene silencing (Eckardt, 2002; Hannon, 2002; Mansoor et al., 2006). The trigger for RNA silencing is the presence of double-stranded RNA (dsRNA), derived either from endogenous or exogenous transcripts or by the production of self-complementary foldback RNAs, which is further cleaved into small interfering RNA (siRNA) or microRNA (miRNA), to become functional in a number of epigenetic gene-silencing processes (Durremberger et al., 1989).

siRNAs are produced from dsRNA by an RNase III termed DICER-like (DCL) in plants, and then incorporated into an RNA-induced silencing complex (RISC), leading to a sequence-specific cleavage of target mRNAs. However, clear differences in siRNA-mediated RNA silencing have also been found between plants and animals; for example, siRNAs in plants fall into two distinct classes: a short (~21 nt) and a long (~24 nt) size class. The 21 nt siRNAs are sufficient to guide the cleavage of target RNA mediated by RISC, while the 24 nt species are probably involved in chromatin modifications (Fusaro et al., 2006; Hamilton et al., 2002; Vaucheret, 2008; Zhang and Zhu, 2011).

RNA silencing also involves RNA-directed DNA methylation (RdDM), in which a DNA, homologous to a dsRNA triggering gene silencing, is methylated de novo (Matzke et al., 2009; Mette et al., 2000; Zhang and Zhu, 2011). Small RNAs guide the RNA-induced transcriptional gene silencing (RITS) complex to direct chromatin modifications and DNA methylation of the homologous DNA sequences (Verdel et al., 2004). RdDM has been documented in many plant systems, and requires the enzyme known as domains rearranged methylase, which could be guided by siRNAs to the target sequences (Cao et al., 2003; Law and Jacobsen, 2010). In contrast to mammals, in which DNA methylation occurs almost exclusively on cytosines in the symmetric dinucleotides CpG, in plants, cytosine methylation occurs both at symmetric sites (CpG and CpNpG, where N is A, T or C) and at asymmetric sites (CpNpN). In plants, several distinct RNA silencing pathways operate to repress gene expression at the transcriptional or post-transcriptional level (Baulcombe, 2004; Law and Jacobsen, 2010). In both cases silencing is generally associated with siRNAs and/or DNA methylation: siRNAs homologous to the promoter region of a target gene induce transcriptional gene silencing (TGS), which is associated with promoter methylation. siRNAs homologous to the coding region of the target gene induce post-transcriptional gene silencing (PTGS), which involves mRNA-specific degradation in the cytoplasm and, in some cases, methylation of the coding sequences (Baulcombe, 2004).

It was proposed that siRNAs of the longer class (24–25 nt) could be the trigger molecules for RdDM (Hamilton et al., 2002). The short class of siRNAs (21 nt) is part of the natural plant defense mechanism against viruses (Voinnet, 2005), and it has been proposed that this class guides the RISC in degrading specific mRNAs in herbaceous plants (Hamilton et al., 2002; Zhang and Zhu, 2011). The accumulation of different siRNA size classes could be the result of different inducers, virus infection, or transgenes. In plants, transgene- and virus-induced RNA silencing pathways are overlapping but not identical (Voinnet, 2005). The virus seems to act as both inducer and target in this process, as observed in grapevine (Gambino et al., 2009, 2010), plum (Hily et al., 2005), peach inoculated with Peach latent mosaic viroid, and chrysanthemum inoculated with *Chrysanthemum chlorotic mottle* viroid (Martinez de Alba et al., 2002). The accumulation of two classes of siRNAs in transgenic *Prunus* resistant to *Plum pox virus* was demonstrated, and it was shown that those siRNAs are constitutively present only in the resistant clone (Hily et al., 2005). The resistance mounted by the shorter siRNAs is apparently overcome and the longer siRNAs are responsible for high levels of resistance against *Plum pox virus*. In transgenic grapevine silenced lines containing T-DNAs in complex arrangements, high levels of transgene methylation were observed in both symmetrical and asymmetrical cytosines in the 35S promoter, GFLV-CP, and T7 terminator sequences (Gambino et al., 2010). These observations confirmed that T-DNA inverted repeats trigger transgene silencing and are susceptible to de novo methylation (Selker, 1999; Zhang and Zhu, 2011). However, the correlation between DNA methylation and multiple copies was not confirmed in all lines, suggesting that transgene silencing in grapevine was not always associated with DNA methylation. In addition, in the GFLV-CP lines, no siRNAs could be detected. The absence of siRNAs, or the inability to identify siRNAs, has also been observed in constructs that should lead to silencing with high efficiency, such as transgenes containing GFLV sequences in hairpin RNA (hpRNA) structures (Carra et al., 2011). However, it is possible that RNA signaling molecules were responsible for the DNA methylation patterns observed in these lines, even if siRNAs were below the detection level. Similar results were observed in *Arabidopsis*, in which the gene *PA1* was methylated at both symmetrical and asymmetrical contexts, but no *PA1* siRNAs could be detected (Melquist and Bender, 2004).

Furthermore, lines showing transgene silencing were unable to prevent the virus spreading simply by graft inoculation (Gambino et al., 2010). After inoculation with GFLV, no change in the levels of cytosine methylation was observed, and both transformed and untransformed plants produced short siRNAs (21–22 nt), indicating that grapevine plants responded to GFLV infection by activating a post-transcriptional gene silencing mechanism. However, these siRNAs were unable to block virus replication, indicating that GFLV could circumvent this silencing pathway. This susceptibility to GFLV could be due to the high viral inoculum and to the constant viral pressure from the rootstock applied to relatively young plants (Gambino et al., 2010). These transgenic grapevines seemed to be unable to suppress GFLV replication. In this case, the silencing process, although effective against the transgene, might be incomplete, thus leaving a small pool of viral transcripts that may be sufficient for GFLV replication.

7.4 Identification and characterization of transgene insertion loci

Since the site of insertion could determine expression levels of transgenes, it is important to know the insertion loci for every transgenic line. An efficient strategy to find the functions of genes in transgenic lines is to sequence the regions flanking T-DNA insertion sites, which also provide a valuable resource for studying the mechanism of T-DNA transformation. T-DNA integration sites have been extensively investigated in the last decade, especially in *Arabidopsis* (Singer et al., 2012; Takano et al., 1997; Tinland, 1996; Windels et al., 2003).

A widely used method to introduce genes into plants is *Agrobacterium tumefaciens*-mediated transformation, which is a unique phenomenon of horizontal gene transfer between prokaryotes and eukaryotes. It involves the processing of T-DNA from the *A. tumefaciens* Ti (tumor-inducing) plasmid and its transfer to the plant (Gelvin, 2000, 2003; Kim et al., 2007). The bacterium transfers a DNA segment (T-DNA) delineated by two imperfect 25 bp direct border repeats, the left border (LB) and the right border (RB) (Gelvin, 2003). Different proposed mechanisms are potentially involved in T-DNA integration (Tzfira et al., 2004; Waterworth et al., 2011). All these models are based on the assumption that no extensive sequence homology exists between T-DNA and the plant chromosome integration site, although micro-homologies (microsimilarity) and/or filler DNA occasionally exist (Brunaud et al., 2002; Gelvin, 2000; Tinland, 1996). Microsimilarities influence the T-DNA integration process and position in the chromosomes, while the filler DNA originates from the repair of heteroduplexes generated from short stretches of non-contiguous micro-homologies (Tzfira et al., 2004; Windels et al., 2003). These microsimilarities may provide a pre-integration docking force for T-DNA strands, influencing the T-DNA integration process and its chromosomal position. However, as a consequence of the short length of microsimilarities, T-DNA might have the potential to integrate in several regions of the plant genome (Gambino et al., 2009). T-DNA integration may therefore occur by a process of non-homologous end-joining and consist of a direct linkage between T-DNA and grapevine genome without deletions of DNA or insertion of filler DNA (Gambino et al., 2009; Gheysen et al., 1991).

Several previous reports, however, have suggested that T-DNA integration into the plant genome might not be completely at random (Kim et al., 2007). Integration of T-DNA in centromeric, non-genic, and other regions has been observed less frequently than would be expected if random integration is assumed (Alonso et al., 2003; Brunaud et al., 2002; Krysan et al., 2002; Szabados et al., 2002). Similarly, in grapevine, a non-random distribution of the T-DNA was observed, with a preference for transcriptionally active regions, and with a bias towards the intron sequences of the genes (Gambino et al., 2009). Two possible explanations have been suggested for the observed non-randomness of mapped T-DNA integration sites (Francis and Spiker, 2005). It is possible that genic regions are preferred sites for T-DNA integration, or that T-DNA integration occurs randomly throughout the entire genome, but integrations into some regions are not recognized due to the repressive nature of the surrounding chromatin or nearby transcriptional regulators. Selection bias is expected to be at least partially responsible for the perceived non-randomness of T-DNA integration. Integration of T-DNA into non-expressed regions of the genome would probably result in plants that would not survive the selection regime. Studies by Dominguez et al. (2002) and Francis and Spiker (2005) conducted under selection-free conditions showed a higher perceived transformation efficiency, corresponding to a higher frequency of transgene silencing. In addition, it was shown by analyzing T-DNA flanking sequences that the frequency of T-DNA integration in transcriptionally inert regions in *Arabidopsis* regenerated without selection is similar to that in transcriptionally active regions (Francis and Spiker, 2005; Kim et al., 2007).

The AT-rich regions are commonly accepted as the preferred sites for T-DNA integration (Gelvin, 2003). Analysis of *Arabidopsis* T-DNA insertion mutant libraries indicates a preference for T-DNA insertion into 5′ gene regulatory regions and AT-rich regions (Brunaud et al., 2002; Kim et al., 2007; Szabados et al., 2002). Compositional analysis of grapevine DNA around the T-DNA insertion sites in transgenic lines (Gambino et al., 2009) revealed the same asymmetric distribution of CG around the insertion sites. AT-rich sequences were present at 300–500 bp upstream of the insertion points near the RBs, while the CG content around and downstream of T-DNA integration sites was close to the average for the grapevine genome. In addition, the asymmetry of CG and AT-skew values at the point of insertion indicates that there was an overabundance of C and T residues upstream of the insertion site, and an overabundance of G and A residues downstream of the insertion site (Gambino et al., 2009). Thus, DNA composition is a likely determinant for the location of T-DNA insertion in grapevine, since nucleotide asymmetries have long been associated with changes in the physical structure of DNA (Brukner et al., 1995).

Altogether, previous studies suggested that the selection pressure might cause a preference for T-DNA (containing the antibiotic resistance genes) integration into transcriptionally active regions of the genome. Integration of T-DNA into non-expressed regions of the genome would probably result in plants that would not survive the selection regime. Studies by Dominguez et al. (2002) and Francis and Spiker (2005) conducted under selection-free conditions showed a higher perceived transformation efficiency, corresponding to a higher frequency of transgene silencing. However, it was shown by analyzing T-DNA flanking sequences that the frequency of T-DNA integration in transcriptionally inert regions in *Arabidopsis* regenerated without selection is similar to that in transcriptionally active regions (Francis and Spiker, 2005; Kim et al., 2007).

7.5 Integration of vector backbone

Although *Agrobacterium* was believed to deliver only the T-DNA sequence between the right and left border sequences, integration of non-T-DNA binary vector backbone sequences into the genome of transgenic plants frequently occurs (Gambino et al., 2009; Kononov et al., 1997; Maghuly et al., 2006c, 2007, 2008). It is known that 25 bp sequences at the right and the left border regions delineate the transferred T-DNA from right to left (Gelvin, 2000). These flanking sequences enhance (on the right border) or attenuate (on the left border) the activity of the T-DNA sequence (Wang et al., 1987). Due to the fact that the flanking sequence of the left border is not strong enough to attenuate the transfer of vector backbone sequences during transformation, backbone DNA contamination may exist commonly in *Agrobacterium*-derived transgenic plants. Vector backbone sequences (VBS) have been found in several dicot (De Buck et al., 2000; Zhang et al., 2008) and also monocot species (Kim et al., 2003; Wu et al., 2006; Zhai et al., 2004). Integration of VBS in transgenic plants is strictly unwanted for crop plants. Regulatory authorities require genetically modified plants to be devoid of unnecessary DNA, especially VBS containing bacterial resistance genes and origins of replication. In grapevines, PCR assays showed that VBS were integrated in 28.6% of the transgenic plants analyzed (Gambino et al., 2009; Maghuly et al., 2006c, 2007). These findings raised considerable concern over genetically modified plants (GMP) because of regulatory requirements by the EU to avoid unnecessary DNA segments and the possible effects of these sequences on gene expression and other functions. This phenomenon requires efficient means of control, as such transgenic plants cannot be subjected to field trial experiments from the year 2008 on, according to EU directive 2001/18/EC (EFSA, 2004; *http://www.efsa.europa.eu/en. efsajournal/pub/48.htm*).

Two mechanisms can explain the transfer of vector backbone sequences into the plant genome: (i) correct initiation at the RB, but inefficient recognition and nicking at the LB repeat, results in read-through vector backbone transfer (De Buck et al., 2000; Yin and Wang, 2000); (ii) the LB repeat can be recognized as an initiation site for T-strand production, because the LB nicking is also associated with VirD2 attachment (Durremberger et al., 1989) to the non-T-DNA portion of the Ti plasmid. In this case, the vector backbone is transferred first to the plant genome (Kim et al., 2003).

The transfer of vector backbone sequences reported by Gambino et al. (2009) is consistent with the hypothesis of initiation of T-DNA transfer at the RB repeat, read-through at the LB and transfer of the complete vector backbone. However, in some lines the transfer of vector sequences might have resulted from T-DNA processing and DNA transfer initiations at the LB repeat. LB nicking could be associated with VirD2 attachment to the non-T-DNA portion of the Ti plasmid, the vector backbone being transferred first to the plant nucleus. In some other lines the vector sequences were not linked to the T-DNA. In some *Vitis* lines the T-DNA and vector backbone were cross-linked between the borders at the 5′ end of the single-strand T-DNAs, suggesting that both ends can act as starting points for transfer. In addition, VBS are linked to integration of multiple T-DNAs in the same position in the grapevine genome, resulting in the formation of tandem and inverted repeats (Gambino et al., 2009). Linkages of different T-DNAs have been observed in other plant species (De Buck et al., 2000; De Neve et al., 1997; Fu et al., 2006; Iglesias et al., 1997; Wenck et al., 1997). It was reported that T-DNAs were frequently integrated at the same locus in all possible configurations but with preference for those with at least one RB involved in the linkage (De Neve et al., 1997). The authors suggested that T-DNA repeats were formed by extrachromosomal ligation of the T-DNAs before integration. Results obtained by Gambino et al. (2009) agree with this model;

however, in lines containing the entire vector backbone sequence some tandem repeats could have been formed at the time of T-strand formation in the bacterial cell.

Analyses by Gambino et al. (2009) indicate that the integration site within the grapevine genome was not responsible for transgene expression level, because generally the T-DNA integrated into transcriptionally competent regions. Transgene silencing in a few lines seems to be associated with integration of multiple transgene copies in tandem or inverted repeats at a single locus, as in some *Vitis* lines (Gambino et al., 2009, 2010). The transfer of VBS seems to be insufficient to induce silencing, but is probably indirectly connected to the transgene expression level, as the transgenic grapevines containing vector backbone sequences show linkages of T-DNAs. Consequently, it is important to produce genetically modified plants containing single or few and minimally rearranged transgene insertions. In the case of grapevines, the use of *Agrobacterium* as the vector benefits from the fact that they are within the host range of *Agrobacterium* and this method of transformation leads to a stable integration into the host DNA, in low copy numbers (Maghuly et al., 2006c, 2007, 2008) as compared with that obtained with non-biological transfer systems, for example, gene gun (Mehlenbacher, 1995). However, these approaches alone will not always overcome problems with the instability of the transgene expression, because transgene silencing has also been observed in cases with single copy insertion (Ainley and Kumpatla, 2004).

7.6 Stability of inserted transgenes

Genetic transformation technology has facilitated studies of gene regulation in several plant species, including perennial trees and grapevines (Aronen et al., 1995; Barampuram and Zhang, 2011; Borejsza-Wysocka et al., 2010; Deikman et al., 1998; Hawkins et al., 2003; Krasnyanski et al., 2001; Maghuly et al., 2007; Newton et al., 1992). Some of the most problematic barriers to the genetic improvement of grapevines, such as their long breeding cycles, can be circumvented by the application of these techniques. Because perennial plants have a long lifespan, knowledge of the genetic regulation of mature tissues is of major importance.

The successful introduction of genetically improved trees and grapevines depends on asserting and improving the agricultural achievement of the modified plants and on the stable expression of the transgene (Laimer, 2007; McElroy, 1999; Tuteja et al., 2012). An understanding of the factors influencing transgene expression and stability is important for three reasons: (i) to gain a better basic understanding of the ways in which genes act and interact; (ii) for assessing the utility of a particular transgene in agriculture; and (iii) for biosafety assessment. Particularly in Europe, potential adverse long-term effects are attracting major attention; among them genetic instability has been repeatedly addressed as a potential hazard (McElroy, 1999). Therefore the need existed for a model fruit species to address questions of behavior and stability of expression of transgenes in woody crop species over a longer observation period. It is important that the transgene is expressed stably in a controlled way. The modifications targeted to all of the plant tissues must continue to be expressed until culmination age, even under stress conditions. In other words, transgene expression should remain stable in time and space throughout the lifetime of the plant (Borejsza-Wysocka et al., 2010; Maghuly et al., 2007). There is less need to consider the inheritance pattern in successive generations in grapevines, since grafting is the normal method of propagating fruit trees, providing unlimited numbers of clones of selected transgenic lines (Ainley and Kumpatla, 2004).

The stability of transgene expression is of vital importance, whenever transgenic plants are produced. Changes in transgene expression levels, often accompanied by changes in DNA methylation levels, have been associated with changes in environmental conditions and stress (Kotakis et al., 2010; Matzke and Matzke, 1995), a situation which could be experienced upon transfer to the greenhouse or the screenhouse (Maghuly et al., 2007). Analysis of methylation in transgenic lines after transfer to greenhouse conditions revealed an increase of symmetric cytosine (CpG) methylation, which could be linked to reduction of transgene expression (Gambino et al., 2010). However, it has been reported that transgene expression in woody plants appears to be stable under in vitro and in vivo conditions, both in the greenhouse and in the screenhouse conditions (Maghuly et al., 2007). It was suggested that gene silencing is relatively rare in woody trees (Fladung et al., 1997), and indeed results obtained by Maghuly et al. (2007) indicate that transgene expression can be stable over long periods under screenhouse conditions, which approximately reflect field conditions. Different expression between different organs and between leaves from different positions along the stem (different stages of maturity) is perhaps associated with the process of lignification, which is more pronounced in woody tissues (Maghuly et al., 2007, 2008; Mahanil et al., 2012). It has also been suggested that the observed differences most probably reflect differences in general metabolic levels (Willmitzer, personal communication). It is known from studies on forest tree species (Ellis et al., 1996) that the activity of the gene of interest is higher under in vitro than under in vivo conditions, and that it is most probably correlated with the physiological activity of the plant material under study (Maghuly et al., 2007). Such results are in agreement with other studies (Brandle et al., 1995; Hawkins et al., 2003; Pinçon et al., 2001), which reported that stress and other changes in environmental conditions, as well as the developmental stage of the plant, could affect the expression level of transgenes and, consequently, the characteristics targeted. In cherries, although the level of transgene expression was reduced, the transgenes continued to be expressed without any case of gene silencing. In addition, phenological observations of the cherry plants during the vegetative growth phase over three consecutive years showed a normal development without morphological differences between the transgenic and the non-transgenic control plants (Maghuly et al., 2007). This is particularly important for trees, which undergo numerous dormancy cycles and are often exposed to extreme environmental changes during their life.

7.7 Conclusions

Even if recent works in grapevine genomics offer promising possibilities for grape breeding, much is still to be done. We have only a very limited knowledge about the genetic background of agronomically important traits. However, structurally and functionally well-annotated reference grape genomes offer a good framework for extensive exome sequencing projects, identification of allele variants and haplotypes of genes of interest, and for exploring the genetic and biological background of desired traits. On the other hand, established grapevine cultivars are an integral part of the economy and cannot be replaced easily by newly bred cultivars. Through the application of directed genetic improvements making new traits available, breeding steps requiring less time and desired traits can be directly introduced into established cultivars.

7.8 Acknowledgement

The financial support of the Austrian Bundesministerium für Wissenschaft und Forschung is acknowledged.

7.9 References

Adam-Blondon AF, Roux C, Claux D, Butterlin G, Merdinoglu D, This P. 2004. Mapping 245 SSR markers on the *Vitis vinifera* genome: a tool for grape genetics. *Theor. Appl. Genetics* 109: 1017–27.

Ainley W, Kumpatla SP. 2004. Gene silencing in plants. In: Parekh SR (Ed.) *The GMO Handbook: Genetically Modified Animals, Microbes and Plants in Biotechnology*. Humana Press Inc., Totowa, NJ. pp. 243–62.

Allison RF, Schneider WL, Greene AE. 1996. Recombination in plants expressing viral transgenes. *Semin. Virol.* 7: 417–22.

Alonso JM, Stepanova AN, Leisse TJ, Kim CJ, Chen HM, et al. 2003. Genome-wide insertional mutagenesis of *Arabidopsis thaliana*. *Science* 301: 653–7.

Aronen T, Hohtola A, Laukkanen H, Häggman H. 1995. Seasonal changes in the transient expression of a 35S CaMV-GUS gene construct introduced into Scots pine buds. *Tree Physiol.* 15: 65–70.

Arroyo-Garcia R, Martinez-Zapater JM. 2004. Development and characterization of new microsatellite markers for grape. *Vitis* 43: 175–8.

Barampuram S, Zhang ZJ. 2011. Recent advances in plant transformation. *Methods Mol. Biol.* 701: 1–35.

Bassi D. 2006. Breeding for resistance: breeding for resistance to *Plum pox virus* in Italy. *OEPP/EPPO Bulletin* 36: 327–9.

Bassi D, Bellini D, Guerriero R, Monastra F, Pennone F. 1995. Apricot breeding in Italy. *Acta Hort.* 384: 47–54.

Battilana J, Constantini L, Emanuelli F, Sevini F, Segala C, et al. 2009. The l-deoxy-D-:-xylulose 5-phosphate synthase gene co-localizes with a major QTL affecting monoterpene content in grapevine. *Theor. Appl. Genetics* 118: 653–69.

Baulcombe D. 2004. RNA silencing in plants. *Nature* 431: 356–63.

Bertioli DJ, Harris DR, Edwards ML, Cooper JI, Hawes WS. 1991. Transgenic plants and insect cells expressing the coat protein of *Arabis mosaic virus* produce empty virus-like particles. *J. Gen. Virol.* 72: 1801–9.

Bhatti S, Jha G. 2010. Current trends and future prospects of biotechnological interventions through tissue culture in apple. *Plant Cell Rep.* 29: 1215–25.

Borejsza-Wysocka E, Norelli JL, Aldwinckle HS, Malnoy M. 2010. Stable expression and phenotypic impact of *attacin E* transgene in orchard grown apple trees over a 12 year period. *BMC Biotechnol.* 10: 41.

Bowers JE, Dangl GS, Vignani R, Meredith CP. 1996. Isolation and characterization of new polymorphic simple sequence repeat loci in grape (*Vitis vinifera* L). *Genome* 39: 628–33.

Brandle JE, McHugh SG, James L, Labbe H, Miki BL. 1995. Instability in transgene expression in field grown tobacco carrying the csrl-1 gene for herbicide resistance. *Nat. Biotechnol.* 13: 994–8.

Brukner I, Sanchez R, Suck D, Pongor S. 1995. Sequence-dependent bending propensity of DNA as revealed by DNaseI: parameters for trinucleotides. *EMBO J.* 14: 1812–18.

Brunaud V, Balzergue S, Dubreucq B, Aubourg S, Samson F, et al. 2002. T-DNA integration into the *Arabidopsis* genome depends on sequences of pre-insertion sites. *EMBO Rep.* 3: 1152–7.

Cabezas JA, Cervera MT, Ruiz-Garcia L, Carreno J, Martinez-Zapater JM. 2006. A genetic analysis of seed and berry weight in grapevine. *Genome* 49: 1572–85.

Cao XF, Aufsatz W, Zilberman D, Mette MF, Huang MS, et al. 2003. Role of the DRM and CMT3 Methyltransferases in RNA-directed DNA methylation. *Curr. Biol.* 13: 2212–17.

Carra A, Gambino G, Urso S, Nervo G. 2011. Non-coding RNAs and gene silencing in grape. In: Erdmann VA, Barciszewski J (Eds) *Non Coding RNAs in Plants*. Springer-Verlag, Berlin. pp. 67–78.

Cipriani G, Spadotto A, Jurman I, Di Gaspero G, Crespan M, Meneghetti S, et al. 2010. The SSR-based molecular profile of 1005 grapevine (*Vitis vinifera* L.) accessions uncovers new synonymy and parentages, and reveals a large admixture amongst varieties of different geographic origin. *Theor. Appl. Genetics* 121: 1569–85.

Costantini L, Battilana J, Lamaj F, Fanizza G, Stella Grando M. 2008. Berry and phenology-related traits in grapevine (*Vitis vinifera* L.): from quantitative trait loci to underlying genes. *BMC Plant Biology* 8: 38.

Dalbó MA, Ye GN, Weeden NF, Wilcox WF, Reisch BI. 2001. Marker-assisted selection for powdery mildew resistance in grapes. *J. Am. Soc. Hortic. Sci.* 126: 83–9.

De Buck S, De Wilde C, Van Montagu M, Depieker A. 2000. T-DNA vector backbone sequences are frequently integrated into the genome of transgenic plants obtained by *Agrobacterium*-mediated transformation. *Mol. Breeding* 6: 459–68.

Decroocq V, Foulongne M, Lambert P, Le Gall O, Mantin C, et al. 2005. Analogues of virus resistance genes map to QTLs for resistance to sharka disease in *Prunus davidiana*. *Mol. Gen. Genomics* 272: 680–9.

Decroocq V, Sicard O, Alamillo JM, Lansac M, Eyquard JP, et al. 2006. Multiple resistance traits control *Plum pox virus* infection in *Arabidopsis thaliana*. *Mol. Plant-Microbe Interact.* 19: 541–9.

Deikman J, Xu R, Kneissl ML, Ciardi JA, Kim KN, Pelah D. 1998. Separation of *cis* elements responsive to ethylene, fruit development, and ripening in the 5'-flanking region of the ripening-related E8 gene. *Plant Mol. Biol.* 37: 1001–11.

De Neve M, De Buck S, Jacobs A, Van Montagu M, Depicker A. 1997. T-DNA integration patterns in co-transformed plant cells suggest that T-DNA repeats originate from co-integration of separate T-DNAs. *Plant J.* 11: 15–29.

Díaz-Riquelme J, Lijavetzky D, Martínez-Zapater JM, Carmona MJ. 2009. Genome-wide analysis of MIKCC-type MADS box genes in grapevine. *Plant Physiol.* 149(1): 354–69.

Di Gaspero G, Cattonaro F. 2010. Application of genomics to grapevine improvement. *Austral. J. Grape Wine Res.* 16: 122–30.

Di Gaspero G, Cipriani G, Marrazzo MT, Andreetta D, Castro MJP, et al. 2005. Isolation of $(AC)_n$-microsatellites in *Vitis vinifera* L and analysis of genetic background in grapevines under marker assisted selection. *Mol. Breeding* 15: 11–20.

Doligez A, Bouquet A, Danglot Y, Lahogue F, Riaz S, et al. 2002. Genetic mapping of grapevine (*Vitis vinifera* L.) applied to the detection of QTLs for seedlessness and berry weight. *Theor. Appl. Genetics* 105: 780–95.

Doligez A, Audiot E, Baumes R, This P. 2006. QTLs for muscat flavor and monoterpenic odorant content in grapevine (*Vitis vinifera* L.). *Mol. Breeding* 18: 109–25.

Dominguez A, Fagoaga C, Navarro L, Moreno P, Pena L. 2002. Regeneration of transgenic citrus plants under nonselective conditions results in high-frequency recovery of plants with silenced transgenes. *Mol. Genet. Genomics* 267: 544–56.

Dondini L, Lain O, Geuna F, Banfi R, Gaiotti F, et al. 2007. Development of a new SSR-based linkage map in apricot and analysis of synteny with existing *Prunus* maps. *Tree Genet. Genomes* 3: 239–49.

Duchêne E, Butterlin G, Claudel P, Dumas V, Jaegli N, Merdinoglu D. 2009. A grapevine (*Vitis vinifera* L.) deoxy-D-xylulose synthase gene colocates with a major quantitative trait loci for terpenol content. *Theor. Appl. Genetics* 118: 541–52.

Durremberger F, Crameri A, Hohn B, Koukolikova-Nicola Z. 1989. Covalently bound VirD2 protein of *Agrobacterium tumefaciens* protects the T-DNA from exonucleolytic degradation. *Proc. Natl. Acad. Sci. U.S.A.* 86: 9154–8.

Eckardt NA. 2002. RNA goes mobile. *Plant Cell* 14: 1433–6.

EFSA 2004. Opinion of the scientific panel on genetically modified organisms on the use of antibiotic resistance genes as marker genes in genetically modified plants (Question No EFSA-Q-2003-109). *EFSA J.* 48: 1–18.

Egea J, Burgos L, Martinez-Gomez P, Dicenta F. 1999. Apricot breeding for sharka resistance at CEBAS-CSIC, Murcia (Spain). *Acta Hort.* 488: 153–7.

Eibach R, Zyprian E, Welter L, Töpfer R. 2007. The use of molecular markers for pyramiding resistance genes in grapevine breeding. *Vitis* 46: 120–4.

Ellis DD, Rintamaki-Strait J, Kleiner K, Raffa K, McCown B. 1996. Transgene expression in spruce and poplar: from the lab to the field. In: Ahuja MR, Boerjan W, Neale DB (Eds) *Somatic Cell Genetics and Molecular Genetics of Trees*. Kluwer Academic Pub., Dordrecht, The Netherlands. pp. 159–63.

Fernandez L, Doligez A, Lopez G, Thomas MR, Bouquet A, Torregrosa L. 2006. Somatic chimerism, genetic inheritance, and mapping of the fleshless berry (*flb*) mutation in grapevine (*Vitis vinifera* L.). *Genome* 49: 721–8.

Fernandez L, Torregrosa L, Segura V, Bouquet A, Martinez-Zapater JM. 2010. Transposon-induced gene activation as a mechanism generating cluster shape somatic variation in grapevine. *Plant J.* 61: 545–57.

Finkel E. 2009. With "phenomics," plant scientists hope to shift breeding into overdrive. *Science* 325: 380.

Fladung M, Kumar S, Ahuja R. 1997. Genetic transformation of Populus genotypes with different chimaeric gene constructs: transformation efficiency and molecular analysis. *Transgenic Res.* 6: 111–21.

Francis KE, Spiker S. 2005. Identification of *Arabidopsis thaliana* transformants without selection reveals a high occurrence of silenced T-DNA integrations. *Plant J.* 41: 464–77.

Fraser RSS. 1990. The genetics of resistance to plant viruses. *Ann. Rev. Phytopathol.* 28: 179–200.

Fu D, Amand PC, Xiao Y, Muthukrishnan S, Liang GH. 2006. Characterization of T-DNA integration in creeping bentgrass. *Plant Sci.* 170: 225–37.

Fuchs M, Gonsalves D. 1995. Resistance of transgenic hybrid squash ZW-20 expressing the coat protein genes of *zucchini yellow mosaic virus* and *watermelon mosaic virus 2* to mixed infections by both potyviruses. *Bio/Technol.* 13: 1466–73.

Fusaro AF, Matthew L, Smith NA, Curtin SJ, Dedic-Hagan J, et al. 2006. RNA interference-inducing hairpin RNAs in plants act through the viral defence pathway. *EMBO Rep.* 7: 1168–75.

Galet P. 2000. *Dictionnaire encyclopédique des cépages*. Hachette Livre, Paris, France.

Gambino G, Gribaudo I, Leopold S, Schartl A, Laimer M. 2005. Molecular characterization of grapevine plants transformed with GFLV resistance genes: I. *Plant Cell Rep.* 24: 655–62.

Gambino G, Chitarra W, Maghuly F, Laimer M, Boccacci P, et al. 2009. Characterization of T-DNA insertions in transgenic grapevines obtained by *Agrobacterium*-mediated transformation. *Mol. Breeding* 24: 305–20.

Gambino G, Perrone I, Carra A, Chitarra W, Boccacci P, et al. 2010. Transgene silencing in grapevines transformed with GFLV resistance genes: analysis of variable expression of transgene, siRNA production and cytosine methylation. *Transgenic Res.* 19: 17–27.

Gelvin SB. 2000. *Agrobacterium* and plant proteins involved in T-DNA transfer and integration. *Annu. Rev. Plant Phys.* 51: 223–56.

Gelvin SB. 2003. *Agrobacterium*-mediated plant transformation: the biology behind the "gene-jockeying" tool. *Microbiol. Mol. Biol. Rev.* 67: 16–37.

Gheysen G, Villarroel R, Van Montagu M. 1991. Illegitimate recombination in plants: a model for T-DNA integration. *Genes Dev.* 5: 287–97.

Golemboski DB, Lomonossoff SF, Zaitlin M. 1990. Plants transformed with a tobacco mosaic virus nonstructural gene sequence are resistant to the virus. *Proc. Natl. Acad. Sci. U.S.A.* 87: 6311–15.

Gölles R, da Câmara Machado A, Minafra A, Savino V, Saldarelli G, et al. 2000. Transgenic grapevines expressing coat protein gene sequences of *grapevine fanleaf virus, arabis mosaic virus, grapevine virus A* and *grapevine virus B*. *Acta Hort.* 528: 305–11.

Goto-Yamamoto N, Mouri H, Azumi M, Edwards KJ. 2006. Development of grape microsatellite markers and microsatellite analysis including oriental cultivars. *Am. J. Enol. Viticult.* 57: 105–8.

Greene AE, Allison RF. 1994. Recombination between viral RNA and transgenic plant transcripts. *Science* 263: 1423–5.

Hamilton A, Voinnet O, Chappell L, Baulcombe D. 2002. Two classes of short interfering RNA in RNA silencing. *EMBO J.* 21: 4671–9.

Hamilton JP, Buell CR. 2012. Advances in plant genome sequencing. *Plant J.* 70: 177–90.

Hannon GJ. 2002. RNA interference. *Nature* 418: 244–51.

Harrison BD, Robinson DJ. 2005. Another quarter century of great progress in understanding the biological properties of plant viruses. *Ann. Appl. Biol.* 146: 15–37.

Hartmann W. 1998. Hypersensitivity: a possibility for breeding sharka resistance plum hybrids. *Acta Hort.* 472: 429–32.

Hartmann W, Neumüller M. 2006. Breeding for resistance: breeding for *Plum pox virus* resistant plums (*Prunus domestica* L.) in Germany. *OEPP/EPPO Bulletin* 36: 332–6.

Hawkins S, Leplé JC, Cornu D, Jouanin L, Pilate G. 2003. Stability of transgene expression in poplar: a model forest tree species. *Ann. For. Sci.* 60: 427–38.

Heffner EL, Sorrells ME, Jannink JL. 2009. Genomic selection for crop improvement. *Crop Science* 49: 1–12.

Hily JM, Scorza R, Webb K, Ravelonandro M. 2005. Accumulation of the long class of siRNA is associated with resistance to *Plum pox virus* in a transgenic woody perennial plum tree. *Mol. Plant-Microbe Interact.* 18: 794–9.

Hoffmann S, Di Gaspero G, Kovács L, Howard S, Kiss S, et al. 2008. Resistance to *Erysiphe necator* in the grapevine 'Kishmish vatkana' is controlled by a single locus through restriction of hyphal growth. *Theor. Appl. Genetics* 116: 427–38.

Iglesias VA, Moscone EA, Neuhuberya F, Michalowski S, Phelan T, et al. 1997. Molecular and cytogenetic analyses of stably and unstably expressed transgene loci in tobacco. *Plant Cell* 9: 1251–64.

Ilardi V, Nicola-Negri ED. 2011. Genetically engineered resistance to *Plum pox virus* infection in herbaceous and stone fruit hosts. *GM Crops* 2: 24–33.

Jacobsen E, Schouten H. 2007. Cisgenesis strongly improves introgression breeding and induced translocation breeding of plants. *Trends Biotechnol.* 25: 219–23.

Jaillon O, Aury JM, Noel B, Policriti A, Clepet C, et al. 2007. The grapevine genome sequence suggests ancestral hexaploidization in major angiosperm phyla. *Nature* 449: 463–8.

Jelly NS, Schellenbaum P, Walter B, Maillot P. 2012. Transient expression of artificial microRNAs targeting Grapevine fanleaf virus and evidence for RNA silencing in grapevine somatic embryos. *Transgenic Res.* 21: 1319–27. DOI: 10.1007/s11248-012-9611-5. (Epub ahead of print)

Khetarpal RK, Maisonneuve B, Maury Y, Chalhoub B, Dinant S, et al. 1998. Breeding for resistance to plant viruses. In: Hadidi A, Khetarpal RH, Koganezawa H. (Eds) *Plant Virus Disease Control*. American Phytopathological Society Press, St Paul. pp. 1–32.

Kim S, Lee J, Jun S, Park S, Kang H, et al. 2003. Transgene structures in T-DNA-inserted rice plants. *Plant Mol. Biol.* 52: 761–73.

Kim S, Veena, Gelvin SB. 2007. Genome-wide analysis of *Agrobacterium* T-DNA integration sites in the *Arabidopsis* genome generated under non-selective conditions. *Plant J.* 51: 779–91.

Kobayashi S, Goto-Yamamoto N, Hirochika H. 2004. Retrotransposon-induced mutations in grape skin color. *Science* 304: 982.

Kononov M, Bassuner B, Gelvin SB. 1997. Integration of T-DNA binary vector "backbone" sequences into the tobacco genome: evidence for multiple complex patterns of integration. *Plant J.* 11: 945–76.

Kotakis C, Vrettos N, Kotsis D, Tsagris M, Kotzabasis K, Kalantidis K. 2010. Light intensity affects RNA silencing of a transgene in *Nicotiana benthamiana* plants. *BMC Plant Biol.* 10: 220.

Krasnyanski SF, Sandhu JS, Osadian MD, Domier LL, Buetow DE, Korban SS. 2001. Effect of an enhanced CaMV 35S promoter and a fruit-specific promoter on uidA gene expression in transgenic tomato plants. *In Vitro Cell Dev. Biol.-Plant* 37: 427–33.

Krysan PJ, Young JC, Jester PJ, Monson S, Copenhaver G, et al. 2002. Characterization of T-DNA insertion sites in *Arabidopsis thaliana* and the implications for saturation mutagenesis. *OMICS* 6: 163–74.

Kuczmog A, Galambos A, Horváth S, Mátai A, Kozma P, et al. 2012. Mapping of crown gall resistance locus *Rcg1* in grapevine. *Theor. Appl. Genet.* Online first. DOI: 10.1007/s00122-012-1935-2.

Laimer M. 2003. The development of transformation of temperate woody fruit crops. In: Laimer M, Rücker W. (Eds) *Plant Tissue Culture: 100 Years Since Gottlieb Haberlandt.* Springer Verlag, Wien. pp. 217–42.

Laimer M. 2007. Transgenic grapevines. *Transgenic Plant J.* 1: 219–27.

Laimer M, Steinkellner H, Weinhäusl A, da Camara Machado A, Katinger H. 1991. Identification of the coat protein gene of Arabis Mosaic Nepovirus and its expression in transgenic plants. *Acta Hort.* 308: 37–41.

Laimer M, Mendonça D, Maghuly F, Marzban G, Leopold S, et al. 2005. Biotechnology of temperate fruit trees. *Acta Biochim. Polonica* 52: 673–8.

Laimer M, Lemaire O, Herrbach E, Goldschmidt V, Minafra A, et al. 2009. Resistance to viruses, phytoplasmas and their vectors in the grapevine in Europe: a review. *J. Plant Pathol.* 91: 7–23.

Laucou V, Lacombe T, Dechesne F, Siret R, Bruno JP, et al. 2011. High throughput analysis of grape genetic diversity as a tool for germplasm collection management. *Theor. Appl. Genet.* 122: 1233–45.

Law JA, Jacobsen SE. 2010. Establishing, maintaining and modifying DNA methylation patterns in plants and animals. *Nat. Rev. Genet.* 11: 204–20. http://www.nature.com/nrg/journal/v11/n3/full/nrg2719.html – a2.

Lijavetzky D, Cabezas JA, Ibáñez A, Rodriguez V, Martínez-Zapater JM. 2007. High throughput SNP discovery and genotyping in grapevine (*Vitis vinifera* L.) by combining a re-sequencing approach and SNPlex technology. *BMC Genomics* 8: 424.

Longstaff M, Brigneti G, Boccard F, Chapman S, Baulcombe D. 1993. Extreme resistance to *potato virus X* infection in plants expressing a modified component of the putative viral replicase. *EMBO J.* 12: 379–86.

Maghuly F, Borroto Fernandez E, Ruthner S, Bisztray G, Pedryc A, Laimer M. 2005. Microsatellite variability in apricots (*Prunus armeniaca* L.) reflects their geographic origin and breeding history. *Tree Genet. Genomes* 1: 151–65.

Maghuly F, Borroto Fernandez E, Zelger R, Marschall K, Katinger H, Laimer M. 2006a. Genetic differentiation of apricot (*Prunus armeniaca* L.) cultivars with SSR markers. *Eur. J. Hort. Sci.* 71: 129–34.

Maghuly F, Borroto Fernandez E, Ruthner S, Pedryc A, Laimer M. 2006b. Microsatellite characterization of apricot (*Prunus armeniaca* L.) cultivars grown in Central Europe. *Acta Hort.* 717: 207–12.

Maghuly F, Leopold S, da Câmara Machado A, Borroto Fernandez E, Khan MA, et al. 2006c. Molecular characterization of grapevine plants with GFLV resistance genes: II. *Plant Cell Rep.* 25: 546–53.

Maghuly F, da Câmara Machado A, Leopold S, Khan MA, Katinger H, Laimer M. 2007. Long-term stability of marker gene expression in *Prunus subhirtella*: a model fruit tree species. *J. Biotechnol.* 127: 310–21.

Maghuly F, Khan MA, Borroto-Fernández EG, Druart P, Watillon B, Laimer M. 2008. Stress regulated expression of the GUS-marker gene (*uid*A) under the control of plant calmodulin and viral 35S promoters in a model fruit tree rootstock: *Prunus incisa x serrula*. *J. Biotechnol.* 135: 105–16.

Mahanil S, Ramming D, Cadle-Davidson M, Owens C, Garris A, et al. 2012. Development of marker sets useful in the early selection of *Ren4* powdery mildew resistance and seedlessness for table and raisin grape breeding. *Theor. Appl. Genet.* 124: 23–33.

Mansoor S, Amin I, Hussain M, Zafar Y, Briddon RW. 2006. Engineering novel traits in plants through RNA interference. *Trends Plant Sci.* 11: 559–65.

Martinez de Alba AE, Flores R, Hernandez C. 2002. Two chloroplastic viroids induce the accumulation of small RNAs associated with posttranscriptional gene silencing. *J. Virol.* 76: 13 094–6.

Matzke MA, Matzke A. 1995. How and why do plants inactivate homologous (Trans) genes? *Plant Physiol.* 107: 679–85.

Matzke M, Kanno T, Daxinger L, Huettel B, Matzke AJ. 2009. RNA-mediated chromatin-based silencing in plants. *Curr. Opin. Cell Biol.* 21: 367–76.

McCarthy MI, Abecasis GR, Cardon LR, Goldstein DB, Little J, et al. 2008. Genome-wide association studies for complex traits: Consensus, uncertainty and challenges. *Nat. Rev. Genet.* 9: 356–69.

McElroy D. 1999. Moving agbiotech downstream. *Nat. Biotechnol.* 17: 1071–4.

McGovern PE. 2003. *Ancient Wine: The Search for the Origins of Viniculture.* Princeton Univ Press, Princeton.

Mehlenbacher SA. 1995. Transformation of woody fruit trees. *Hort. Science* 30: 466–77.

Melquist S, Bender J. 2004. An internal rearrangement in an *Arabidopsis* inverted repeat locus impairs DNA methylation triggered by the locus. *Genetics* 16: 437–48.

Mendel G. 1866. Versuche über Pflanzenhybriden. *Verhandlungen des Naturforschenden Vereines in Brünn* 4: 3–47.

Merdinoglu D, Butterlin G, Bevilacqua L, Chiquet V, Adam-Blondon AF, Decroocq S. 2005. Development and characterization of a large set of microsatellite markers in grapevine (*Vitis vinifera* L) suitable for multiplex PCR. *Mol. Breeding* 15: 349–66.

Mette MF, Aufsatz W, van der Winden J, Matzke MA, Matzke AJM. 2000. Transcriptional silencing and promoter methylation triggered by double-stranded RNA. *EMBO J.* 19: 5194–201.

Moroldo M, Paillard S, Marconi R, Fabrice L, Canaguier A, et al. 2008. A physical map of the heterozygous grapevine 'Cabernet Sauvignon' allows mapping candidate genes for disease resistance. *BMC Plant Biol.* 8: 66.

Myles S, Chia JM, Hurwitz B, Simon C, Zhong GY, et al. 2010. Rapid genomic characterization of the genus *Vitis*. *PLoS ONE* 5: e8219.

Myles S, Boyko AR, Owens CL, Brown PG, Grassi F, et al. 2011. Genetic structure and domestication history of the grape. *Proc. Natl. Acad. Sci. U.S.A.* 108: 3530–5.

Nejidat A, Beachy RN. 1990. Transgenic tobacco plants expressing a coat protein gene of *tobacco mosaic virus* are resistant to some other *tobamoviruses*. *Mol. Plant-Microbe Interact.* 3: 247–51.

Nelson RS, Powell AP, Beachy RN. 1987. Lesions and virus accumulation in inoculated transgenic tobacco plants expressing the coat protein gene of *tobacco mosaic virus*. *Virology* 158: 126–32.

Newton RJ, Yibrah HS, Dong N, Clapham DH, von Arnold S. 1992. Expression of an abscisic acid promoter in *Picea abies* (L.) Krast. following bombardment from an electric discharge particle accelerator. *Plant Cell Rep.* 11: 188–91.

Ooms G, Bakker A, Molendijk L, Wullems GJ, Gordon MP, et al. 1982. T-DNA organization in homogeneous and heterogeneous octopyne-type crown gall tissues of *Nicotiana tabacum*. *Cell* 30: 589–97.

Pantaleo V, Szittya G, Moxon S, Miozzi L, Moulton V, et al. 2010. Identification of grapevine microRNAs and their targets using high-throughput sequencing and degradome analysis. *Plant J.* 62: 960–76.

Pinçon G, Chabannes M, Lapierre C, Pollet B, Ruel K, et al. 2001. Simultaneous down-regulation of caffeic/5-hydroxy ferulic acid-O-methyltransferase I and cinnamoyl coenzyme A reductase in the progeny from a cross between tobacco lines homozygous for each transgene: consequences for plant development and lignin synthesis. *Plant Physiol.* 126: 145–55.

Powell AP, Nelson RS, De B, Hoffman N, Rogers SG, et al. 1986. Delay of disease development in transgenic plants that express the *tobacco mosaic virus* coat protein gene. *Science* 232: 738–43.

Prins M, Laimer M, Noris E, Shubert J, Wassenegger M, Tepfer M. 2008. Strategies for antiviral resistance in transgenic plants. *Mol. Plant Pathol.* 9: 73–83.

Reisch BI, et al. 2012. Grape. In: Badenes ML, Byrne DH (Eds) *Fruit Breeding*. Springer, New York. pp. 225–62.

Riaz S, Tenscher AC, Rubin J, Graziani R, Pao SS, Walker MA. 2008. Fine-scale genetic mapping of two Pierce's disease resistance loci and a major segregation distortion region on chromosome 14 of grape. *Theor. Appl. Genet.* 117: 671–81.

Robinson DJ. 1996. Environment risk assessment of releases of transgenic plants carrying virus-derived inserts. *Transgen. Res.* 5: 359–562.

Safarnejad MR, Jouzani GS, Tabatabael M, Twyman RM, Schillberg S. 2011. Antibody-mediated resistance against plant pathogens. *Biotechnol. Adv.* 29: 961–71.

Sanford JC, Johnston SA. 1985. The concept of parasite derived resistance deriving resistance genes from the parasite's own genome. *J. Theor. Biol.* 113: 395–405.

Sefc KM, et al. 1999. Identification of microsatellite sequences in *Vitis riparia* and their applicability for genotyping of different *Vitis* species. *Genome* 42: 367–73.

Selker EU. 1999. Gene silencing: Repeats that count. *Cell* 97: 157–60.

Singer K, Shiboleth YM, Li J, Tzfira T. 2012. Formation of complex extrachromosomal T-DNA structures in *Agrobacterium*-infected plants. *Plant Physiol.* Preview. DOI: 10.1104/pp.112.200212.

Spielmann A, Krastanova S, Douet-Orhant V, Gugerli P. 2000. Analysis of transgenic grapevine (*Vitis rupestris*) and *Nicotiana benthamiana* plants expressing an *Arabis mosaic virus* coat protein gene. *Plant Sci.* 156: 235–44.

Szabados L, Kovács I, Oberschall A, Abrahám E, Kerekes I, et al. 2002. Distribution of 1000 sequenced T-DNA tags in the *Arabidopsis* genome. *Plant J.* 32: 233–42.

Takano M, Egawa H, Ikeda JE, Wakasa K. 1997. The structures of integration sites in transgenic rice. *Plant J.* 11: 353–61.

Tenllado F, Garcia-Luque I, Serra MT, Diaz-Ruiz JR. 1996. Resistance to pepper mild mottle *tobamovirus* conferred by the 54-KDa gene sequence in transgenic plants does not require expression of the wild type 54-KDa protein. *Virology* 219: 330–5.

Tepfer M. 1993. Viral genes and transgenic plants: what are the potential environmental risks? *Bio/Technol.* 11: 1125–32.

Tester M, Langridge P. 2010. Breeding technologies to increase crop production in a changing world. *Science* 327: 818–22.

This P, Jung A, Boccacci P, Borrego J, Botta R, et al. 2004. Development of a standard set of microsatellite reference alleles for identification of grape cultivars. *Theor. Appl. Genet.* 109: 1148–58.

This P, Lacombe T, Thomas MR. 2006. Historical origins and genetic diversity of wine grapes. *Trends Genet.* 22: 511–19.

Thomas MR, Matsumoto S, Cain P, Scott NS. 1993. Repetitive DNA of grapevine: classes present and sequences suitable for cultivar identification. *Theor. Appl. Genet.* 86: 173–80.

Thomas MR, Cain P, Scott NS. 1994. DNA typing of grapevines: a universal methodology and database for describing cultivars and evaluating genetic relatedness. *Plant Mol. Biol.* 25: 939–49.

Tinland B. 1996. The integration of T-DNA into plant genomes. *Trends Plant Sci.* 1: 178–84.

Tricoli DM, Carney KJ, Russell PF, McMaster JR, Groff DW, et al. 1995. Field evaluation of transgenic squash containing single or multiple virus coat protein gene constructs for resistance to *cucumber mosaic virus*, *watermelon mosaic virus 2* and *zucchini yellow mosaic virus*. *Bio/Technol.* 13: 1458–65.

Troggio, M, Malacarne G, Coppola G, Segala C, Cartwright DA, et al. 2007. A dense single-nucleotide polymorphism-based genetic linkage map of grapevine (*Vitis vinifera* L.) anchoring Pinot Noir bacterial artificial chromosome contigs. *Genetics* 176: 2637–50.

Tuteja N, Verma S, Sahoo RK, Raveendar S, Reddy IN. 2012. Recent advances in development of marker-free transgenic plants: regulation and biosafety concern. *J. Biosci.* 37: 167–97.

Tzfira T, Li J, Lacroix B, Citovsky V. 2004. *Agrobacterium* T-DNA integration: molecules and models. *Trends Genet.* 20: 375–83.

Valat L, Fuchs M, Burrus M. 2006. Transgenic grapevine rootstock clones expressing the coat protein or movement protein genes of grapevine fanleaf virus: characterization and reaction to virus infection upon protoplast electroporation. *Plant Sci.* 170: 739–47.

Vaucheret H. 2008. Plant ARGONAUTES. *Trends Plant Sci.* 13: 350–58.

Velasco R, Zharkikh A, Troggio M, Cartwright DA, Cestaro A. 2007. A high quality draft consensus sequence of the genome of a heterozygous grapevine variety. *PLoS ONE* 2: e1326.

Verdel A, Jia S, Gerber S, Sugiyama T, Gygi S, et al. 2004. RNAi-mediated targeting of heterochromatin by the RITS complex. *Science* 303: 672–6.

Vigne E, Komar V, Fuchs M. 2004. Field safety assessment of recombination in transgenic grapevines expressing the coat protein gene of *Grapevine fanleaf virus*. *Transgenic Res.* 13: 165–79.

Voinnet O. 2005. Induction and suppression of RNA silencing: insights from viral infections. *Nat. Rev. Genet.* 6: 206–20.

Wally O, Punja ZK. 2010. Enhanced disease resistance in transgenic carrot (Daucus carota L.) plants over-expressing a rice cationic peroxidase. *Planta* 232: 1229–39.

Wang K, Genetello C, van Montagu M, Zambryski P. 1987. Sequence context of the T-DNA border repeat elements determines its relative activity during T-DNA transfer to plant cells. *Mol. Gen. Genet.* 210: 338–46.

Waterhouse PM, Graham MW, Wang MB. 1998. Virus resistance and gene silencing in plants is induced by double-stranded RNA. *Proc. Natl. Acad. Sci. U.S.A.* 95: 13959–64.

Waterworth WM, Drury GE, Bray CM, West CE. 2011. Repairing breaks in the plant genome: the importance of keeping it together. *New Phytologist* 192: 805–22.

Welter LJ, Göktürk-Baydar N, Akkurt M, Maul E, Eibach R, et al. 2007. Genetic mapping and localization of quantitative trait loci affecting fungal disease resistance and leaf morphology in grapevine (*Vitis vinifera* L). *Mol. Breeding* 20: 359–74.

Wenck A, Czako M, Kanevsky I, Marton L. 1997. Frequent collinear long transfer of DNA inclusive of the whole binary vector during *Agrobacterium*-mediated transformation. *Plant Mol. Biol.* 34: 913–22.

Windels P, De Buck S, Van Bockstaele E, De Loose M, Depicker A. 2003. T-DNA integration in *Arabidopsis* chromosomes. Presence and origin of filler DNA sequences. *Plant Physiol.* 133: 2061–8.

Wu H, Sparks CA, Jones HD. 2006. Characterisation of T-DNA loci and vector backbone sequences in transgenic wheat produced by *Agrobacterium*-mediated transformation. *Mol. Breeding* 18: 195–208.

Xu K, Riaz S, Roncoroni NC, Jin Y, Hu R, et al. 2008. Genetic and QTL analysis of resistance to *Xiphinema index* in a grapevine cross. *Theor. Appl. Genet.* 116: 305–11.

Yin Z, Wang GL. 2000. Evidence of multiple complex patterns of T-DNA integration into the rice genome. *Theor. Appl. Genet.* 100: 461–70.

Zhai W, Chen C, Zhu X, Chen X, Zhang D, et al. 2004. Analysis of T-DNA-*Xa21* loci and bacterial blight resistance effects of the transgene *Xa21* in transgenic rice. *Theor. Appl. Genet.* 109: 534–42.

Zhang H, Zhu JK. 2011. RNA-directed DNA methylation. *Curr. Opin. Plant Biol.* 14: 142–47.

Zhang J, Cai L, Cheng J, Mao H, Fan X, et al. 2008. Transgene integration and organization in Cotton (*Gossypium hirsutum* L.) genome. *Transgenic Res.* 17: 293–306.

Zohary D, Spiegel-Roy P. 1975. Beginnings of fruit growing in the Old World. *Science* 187: 319–27.

Zohary D, Hopf M. 1993. Fruit trees and nuts. In: Zohary D, Hopf M (Eds) *Domestication of Plants in the Old World*. Oxford Science Publ., Oxford. pp. 134–80.

8

Grapevine genomics and phenotypic diversity of bud sports, varieties and wild relatives

Gabriele Di Gaspero and Raffaele Testolin,
Dipartimento di Scienze Agrarie e Ambientali,
University of Udine, Italy, and Institute of Applied
Genomics/Istituto di Genomica Applicata, Parco
Scientifico e Tecnologico Luigi Danieli, Italy

DOI: 10.1533/9781908818478.149

Abstract: Grapevine (*Vitis vinifera* L.) encompasses a staggering complexity of biochemical variation in the fruit of varieties that individually differ by nuances of colour, taste and aroma. A deep understanding of the underlying DNA variation unlocks the evolutionary relationships between varieties, their somatic mutants and their living wild relatives, and offers insights into the causes of phenotypic diversity. The year 2012 marks the fifth anniversary of the publication of the grapevine genome sequence. The available genomic resources and the new generation of sequencing technologies enable viticulturists to assess for the first time the network of genetic relationships among grapevines on a global scale of genome-wide and whole-population level.

Key words: NGS technologies, microsatellites, transposable elements (TE), TE insertion, structural variations, copy number variation (CNV), linkage disequilibrium (LD), secondary metabolism, transcription factors, gene duplications, bud organogenesis, phenotype differences.

8.1 Introduction

The biological events that gave rise to the centuries-old varieties that still lead the market today have always intrigued historians and viticulturists. The genealogy of even the most famous varieties remained obscure until the turn of the millennium, when DNA analysis supported or falsified common beliefs and historical records, by establishing proof of kinship and pedigrees on a solid scientific basis.

The discovery of the family history of the renowned varieties Cabernet Sauvignon, Chardonnay, Merlot and Muscat of Alexandria revealed that all of them were the result of crosses between other existing varieties (Bowers and Meredith 1997; Bowers et al. 1999; Boursiquot et al. 2009; Cipriani et al. 2010). The ancestors one generation back from these seedlings were themselves famous varieties (for example, the parents of Cabernet Sauvignon are Cabernet franc and Sauvignon blanc), or one celebrated (Pinot noir) and the other neglected (Gouais blanc) for wine quality (in the case of Chardonnay), or one popular and the other rare, as illustrated by Merlot's parents Cabernet franc and Magdeleine Noire des Charentes, and Muscat of Alexandria's parents Muscat à petits grains and Axina de tres bias.

The analysis of nuclear microsatellites proved that close kin varieties are much more widespread than common sense would suggest. Despite many claims for the existence of rich phenotypic diversity in grape genetic resources of Western Europe, most of the varieties analysed by DNA fingerprinting are interconnected by first-degree relationships, regardless of the current geographical area of cultivation.

Beyond parentages, the structure of genetic diversity in grape germplasm has yet to be clarified. The population is weakly structured once the effects of familial relatedness are accounted for (Aradhya et al. 2003). Some of this structure is geographical, because the crop moved from the Near East into Europe, there mating with locally adapted wild grapevines (Myles et al. 2011). More important is the genetic structure associated with the end-use, that is, fruit consumption versus wine production (Myles et al. 2011). In ancient times, table grapes were selected for elongated and branching clusters with large and firm-fleshed berries, the opposite of wine grapes. The structure among wine grapes is extremely weak, albeit detectable, at lower geographical levels (country, wine region), as a result of the human introduction of genetically unrelated varieties, which caused extensive admixture (Cipriani et al. 2010).

The effects of domestication, early selection of varieties in the cradle of viticulture in ancient times, and the constraints imposed by the effect of these human activities on modern varieties are obscure. The available technology before the advent of parallel sequencing allowed only the use of DNA fragments and haplotypes of a few nuclear genes controlling viticultural traits for investigations into grapevine evolution. Despite this limitation, chloroplast DNA gave sufficient insight into the historical changes in the geographical distribution of the different lineages. Without the compounding effect of nuclear DNA contributed by pollen flow, it was established in *Vitis vinifera* that four chlorotypes are ancestral and all are represented in cultivated and wild grapevines in a restricted region in the southern Caucasus (Pipia et al. 2012) – the cradle of viticulture (Olmo 1996; Zohary 1996). Another four chlorotypes are derived and are much more rare (fewer than 5 per cent), but they are all present in wild grapevines dispersed across the Near East in the same tiny corner of the world (Arroyo-García et al. 2006). This observation has been interpreted as an evidence for a single major center of origin.

In this chapter, we review the current knowledge on the structure of genetic diversity in *Vitis vinifera* and the prospect of novel approaches to DNA sequencing, initially called next-generation sequencing (NGS), being routinely applied to germplasm management in viticulture. NGS-enabled methods are improving our ability to address biological questions of speciation, domestication, selective sweeps, and haplotype evolution. In viticulture, this is expected to better clarify the evolutionary history of a highly heterogeneous germplasm and the legacy of ancient wine regions in the birth of modern varieties.

8.2 Origin of *Vitis vinifera*, domestication, and early selection for fruit characters

The ancestral lineages that gave rise to the species *V. vinifera* were introduced from the Far East (Péros et al. 2010). Before human intervention, the grapevine genetic history in western Eurasia was dominated by reduction in diversity and genetic drift, as wild vines retreated during the Ice Ages into isolated refuges in warmer riparian, coastal, and insular areas of the Mediterranean basin and on the southern slopes of the Pyrenees, the Alps and the Caucasus.

The process of domestication made little change to the architecture and biology of the grape plant, and imposed a weak genetic bottleneck (This et al. 2006; Myles et al. 2011). The reduction in haplotype diversity in the domesticated population is statistically significant, but negligible at a genome-wide scale. The rapid decay in linkage disequilibrium (LD), which is observed to a similar extent in the wild and the cultivated compartments, is also inconsistent with the expected effect of major bottlenecks, which would increase LD and homogenise minor allele frequencies in the crop compared with the wild population (Hamblin et al. 2011).

The most conspicuous domestication traits are hermaphroditism, fruitfulness and elongated seed shape. Functional hermaphroditism was the most sought-after trait for making the grape plant worth the labour of cultivation. However, the flower sex locus is genetically identified in the upper arm of chromosome 2, but haplotype diversity in cultivated varieties is not significantly reduced relative to haplotype diversity in wild grapevines, at the level of resolution provided by state-of-the-art SNP chips (Myles et al. 2011). On the other hand, a strongly reduced haplotype diversity in cultivated varieties versus wild relatives is found on chromosome 17, though it remains unknown which genes, function and trait underlie this domestication syndrome.

Once grapevine cultivation became established, different varieties were sought to differentiate fruit flavour, colour, texture and shape. Haplotypic structure and drop of haplotype diversity are observed when rare alleles for favourable traits were strongly selected and fixed. White fruiting grapevines display reduced diversity at a locus of MybA transcript factors and increased LD compared with ancestral red varieties, leading to extensive haplotype structure (Fournier-Level et al. 2010; Myles et al. 2011). The white-skin allele at the MybA locus, which impairs anthocyanin biosynthesis in berry epidermis, has not been found in wild populations that invariantly accumulate anthocyanins, but was favoured in the crop and subject to a strong positive selection. There is significant genetic differentiation within V. vinifera between wine and table grapes; the two types of grapes are the result of bidirectional selection for fruit-related characters, such as berry size, seed dry matter and firmness of flesh texture, that are all negatively correlated with wine quality attributes, but desirable for fresh consumption.

The practice of vegetative propagation became compulsory to maintain the identity of heterozygous varieties, but it has favoured the accumulation of somatic mutations. During many cycles of bud organogenesis to obtain copies of the original stock, some mutations caused noticeable phenotypic effects and created diversity within varieties. These natural variants were the target of clonal selection. In wine regions where restrictive regulations define the list of varieties allowed for cultivation, somatic mutation is the only source of phenotypic variation that generates diversity without modifying the identity of the variety.

8.3 Sources of phenotypic variation in present-day grapevines

Today, the global population of *Vitis vinifera* in Europe, North Africa and the Near East is a mixture of:

- purely wild native grapevines, locally dispersed in small and isolated wild populations (Levadoux 1956)
- cultivated varieties originally domesticated in the Near East, and then dispersed by human trade into other viticulture regions
- seedlings selected from either accidental or intentional hybridisation between introduced varieties
- selections of naturalised grapevines such as introgressive hybrids between varieties and local wild forms (Di Vecchi-Staraz et al. 2009), between wild forms and rootstock material escaped from vineyards (wild American grapevine species) and between cultivated varieties and rootstock material (Arrigo and Arnold 2007)
- de novo domesticated forms from indigenous wild grapevines, though the experimental data supporting this occurrence are still controversial (reviewed in De Andrés et al. 2012)
- somatic mutants of cultivated varieties.

This genetic heterogeneity in local germplasm compels the identification of true autochthonous grapevines that are regarded as an essential component of wine typicity – the degree to which (i) a wine reflects the signature characteristics of the grape variety from which it was produced; and (ii) the variety does well in the soil and climate where it was historically grown. According to general opinion in the European wine market, this link between varieties and wine regions – referred to as terroir by wine magazines and as genotype × environment interaction by plant scientists – would lay the ground for the superiority of traditional wines.

8.4 Genomic tools in the genome sequencing era

The nuclear genome of a model grapevine has been entirely assembled and decoded (Jaillon et al. 2007). The release of the reference sequence in 2007 provided surprising insights into the paleopolyploidy of plant genomes, and also offered a framework against which to compare any other variety in parallel analyses of genome-wide polymorphisms. DNA short-reads are generated from other grapevines, using one of the available instrument platforms for NGS, and aligned against the assembled reference sequence (Mardis 2011). Bioinformatics algorithms and pipelines are then utilised for calling single nucleotide polymorphisms (SNPs), insertions/deletions (indels), and structural variants of individual genomes. At the time the first NGS analysis was performed on grapevine (Myles et al. 2010), DNA reads were no longer than 36 bp and genomic libraries were a sample of a reduced portion of the genome, compounding the complexity of sequence alignment and obscuring to some extent the heterozygosity of grapevines. Despite these limitations, ten times as many SNPs could be identified – and then validated in unrelated varieties beyond those used for SNP discovery – as in the more expensive Sanger sequencing of the single heterozygous variety Pinot noir (Pindo et al. 2008; Vezzulli et al. 2008).

The first attempt to describe sequence diversity in the entire grapevine genome made use of microarrays with oligos that targeted 9k SNPs (Myles et al. 2011), previously identified in a set of ten grape varieties (Myles et al. 2010). Patterns of global diversity revealed genetic footprints of a major domestication event in the Near East, the spread of ancient varieties to the shores of the Mediterranean Sea and inland in Central Europe, and hybridisation with wild forms during the westward expansion. The fixation index F_{ST} also suggested minimal genetic differentiation between populations (Aradhya et al. 2003; Myles et al. 2010).

8.5 Current activities in grapevine genome analysis

The reconstruction of the complete DNA sequence from a model individual was a milestone in the discovery of the composition and structure of the grapevine genome, but a single genome is insufficient to assess the biological diversity of the species.

Genome resequencing (*http://www.vitaceae.org/index.php/Current_Sequencing_Projects*) is now advancing at an increasing pace.

Many resequencing projects are yet to release their results into the public domain. By contrast, past research in molecular genetics showed more success with fewer resources, in terms of rapid translation into practical applications. Molecular genetic tools have received immediate attention from grapevine breeders (reviewed in Di Gaspero and Cattonaro 2010), while the wealth of DNA information produced by NGS is rarely available in its entirety to the scientific community before data validation and interpretation are complete. The high initial investment required for establishing the NGS technology forced many research institutions to adopt a restrictive disclosure policy until all relevant biological information could be extracted and translated into publications. The NGS raw data are also computationally unmanageable by the storage and processing units commonly available in a computer laboratory, and obscure to an end-user like a breeder.

The launch of a 20k SNP chip is the first expected outcome of a community effort of resequencing, involving the Grape ReSeq Consortium and the Institute of Applied Genomics, in collaboration with the technology provider Illumina. In the Illumina 20k SNP chip, three-quarters of the beads are targeted to SNPs identified in a large and highly diverse set of varieties and their wild relatives, and the remaining quarter to SNPs identified in other *Vitis* species. SNPs are balanced for allele frequency, the majority with a minor allele frequency (MAF) above 0.1 in the discovery set – in order to maximise the chance of the SNP to be widely informative in the germplasm – and a small minority with MAF > 0.05 to explore rare alleles. Compared with available genomic tools, this advanced chip gives enough flexibility to explore more comprehensively the genetic diversity in the entire gene pool and to support breeding programs that manipulate traits from wild species.

8.6 Bud organogenesis, somatic mutations, and DNA typing of somatic chimeras

Historically, once a superior grapevine seedling was selected, vegetative propagation then secured the maintenance of genetic consistency and varietal identity over time. A novel grape variety may only originate from sexual reproduction, while somatic mutations that occur in meristematic cells give rise to phenotypic variants within a variety, fixed and perpetuated by vegetative propagation. While biologists may have this concept of variety and somatic clones squared away in their minds, the common sense of viticulturists was challenged in the past by borderline cases.

Somatic variants may phenotypically diverge from the mother plant so extremely that the vegetative material propagated thereafter was mistakenly considered a distinct variety, as opposed to the vegetative material that conserved the characteristics of the original stock. This occurred to many bud sports of the glabrous-leafed, black-waxed and compact-berried Pinot noir that were elevated to the rank of variety, such as the yellowish-berried Pinot blanc, the red-greyish-berried Pinot gris, the hairy-leafed Pinot Meunier, the unwaxed-berried Pinot moure, and the loose-berried Mariafeld types of Pinot noir.

In many cases mutations produce less pronounced phenotypic effects, the resulting variation still staying in the range of the original variety, but sufficient for the new clonal material to be uniquely distinguished, patented and traded with a distinct clone name. Most mutations do not lead to phenotypic variants, and this represents the hidden genetic variation within varieties. The rate of somatic mutations on a genome-wide scale is completely unknown. It is assumed that the older the original seedling and the wider the acreage of planting, the higher the chance that present-day vines are the result of independent and/or sequential somatic mutations. It is also unknown whether some genotypes or specific genome regions are more prone to accumulate somatic mutations. DNA typing of plant somatic mutants is as challenging as sequencing cancer genomes in human tumour tissues. Any tissue section normally selected for genomic DNA isolation (leaf, berry, flower) will include normal cells and mutated cells in unpredictable proportions, which causes overlapping between normal DNA signatures and altered signatures provided by the population of mutated cells.

8.7 Phenotypically divergent clones and the underlying DNA variation

Somatic mutations are sequestered in stratified layers of the apical meristem, giving rise to populations of cells that genetically differ in the same individual – a phenomenon known as somatic chimerism or mosaicism. In these layers, genetic alterations are caused by DNA polymerase slippage, erroneous DNA replication and repair, transposition of mobile elements, and DNA double-strand break followed by non-allelic homologous recombination or non-homologous end joining. Such changes can directly influence phenotypic characteristics, if they alter coding regions or regulatory elements, but they can also affect epigenetic patterns and levels of gene expression without modifying gene sequences. There exist illuminating examples in various grape varieties.

Microsatellites are hyper-variable DNA stretches prone to DNA polymerase slippage. Occasionally, they are variable between cell populations in somatic mosaics and between clones. The interrogation of fewer than a hundred microsatellites was enough to disclose DNA variation among some, but not all, clones of Cabernet Sauvignon and Pinot noir (Hocquigny et al. 2004; Moncada et al. 2006). This variation occurred in intergenic regions and was not associated with phenotypic diversity.

By contrast, a single nucleotide mutation that occurred in the outer cell layer of the meristem and caused an amino acid substitution in the DELLA domain of the gibberellic acid-signalling protein GAI1 is the DNA variant that differentiates the hairy-leafed Pinot Meunier from the original genome of Pinot noir. When mutated cells were isolated from the chimeric apex of Pinot Meunier and used for somatic embryogenesis, the regenerated non-chimeric *GAI1* mutants are dwarf and hyper-fruity, as a result of impaired gibberellic acid signalling and abnormal floral induction opposite every leaf (Boss and Thomas 2002).

White fruiting varieties repeatedly give rise to bud sports that accumulate anthocyanins in their berry skin. Mutations are caused by independent events of partial excision of the Class I long terminal repeat (LTR)-retrotransposon *Gret1* from the promoter region of the MybA1 transcription factor by intra-LTR recombination, which partially restores expression of the transcription factor, thereby leading to the synthesis of the key enzyme for anthocyanin biosynthesis (reviewed in Pelsy 2010).

Transposon-mediated *cis*-activation of a gene homologous to the *Arabidopsis* TERMINAL FLOWER 1 (TFL1) – involved in inflorescence development – is responsible for the reiterated reproductive meristem (RRM) phenotype of the RRM somatic mutant of Carignan. In that case, the insertion of a class II DNA transposon in the promoter of *TFL1* enhances *TFL1* expression and causes exaggerated proliferation and branching of the inflorescence, resulting in huge clusters, along with minor alterations in flower morphology and delayed anthesis (Fernandez et al. 2010).

A structural deletion of a block of DNA amounting to 260 kb is responsible for the complete elimination of the MybA gene cluster from the once red allele of Pinot noir in its derived white fruiting mutant Pinot blanc (Yakushiji et al. 2006).

In addition to these well-documented cases, observations of somatic variation of human cells provide arguments that clonal variants in plants may also arise from copy number variation, epigenetic modifications, and misregulation of microRNA and small RNA pathways, though experimental evidence for these mechanisms has yet to come in grapevine.

8.8 Transposon insertion-site profiling using NGS

Most of the known phenotypic variation between grapevine clones is accounted for by the activity of transposable elements (TE) (Benjak et al. 2009). This evidence emerged from the investigation of gene regions already known to be responsible for trait variation (i.e. *MybA*) or transcriptionally altered in mutants (i.e. *TFL1*). This generated the expectation that transposition of Class I and II mobile elements is the most frequent cause of somatic mutations.

Systematic monitoring of transposon activity is today possible and relatively easy with the use of NGS. The earliest application of NGS to address the question of clonal variation made use of gapped alignment of Roche 454 GS FLX single reads, averaging 355 bp in length, from clones of Pinot noir (Carrier et al. 2012). The first estimates should be taken with caution due to low coverage sequencing, but they indicate a mutation rate of 35 TE polymorphic sites per million nucleotides among clones, compared with as few as 1.6 SNPs and 5.1 small indels per million nucleotides. Most of this variation is thought to be phenotypically silent, but more than half of those events were localised in genic regions, which may postulate some biological relevance worth testing. Shorter but paired sequences are now available from both ends of sheared and size-selected fragments using Illumina and Applied Biosystems NGS platforms. Reads of this sort can be handled by diverse bioinformatic tools to scan the genome for deletions and novel (non-reference) TE insertions: (i) gapped alignment of reads or so-called split-read method; (ii) deviation from mean library insert-size in paired-end mapping; and (iii) mapping to a single genome region of two groups of unpaired reads orphaned by non-alignment of their mates, which are then assembled into sequence contigs and compared with TE databases. A genome-wide transposon insertion-site profiling may become the method of choice for the systematic scanning of somatic variation.

8.9 Large structural variation using NGS

Grapevine homologous chromosomes differ in grapevine varieties by non-colinear DNA stretches located in orthologous positions, amounting to an estimated 65 million nucleotides, and by the presence/absence of another 49 million nucleotides (Velasco et al. 2007). Some of these structural variants are longer in size than TEs and involve dispensable genes. The example given above for the berry colour locus in Pinot noir and Pinot blanc indicates that a similar source of genetic variation also arises from somatic mutations, may affect hundreds of kilobases, and may generate phenotypic differences between clones by changing their gene content (Yakushiji et al. 2006). Large structural variation and copy number variants are usually scanned through an entire genome by using array-based technologies, with an approach called comparative genome hybridisation, or by mapping NGS reads and detecting aberrant depth of coverage. The latest technical advances in library preparation now allow sequencing of ends from fosmid-size fragments (Williams et al. 2012). Fosmids have consistent and narrow insert-size distribution averaging around 35–40 kb. The large spans of their mapped paired-ends facilitate the detection of mobile element transposition as well as larger structural variations.

8.10 Copy number variation, gene redundancy, and subtle specialisation in secondary metabolism

Phenotypic diversity was expected to stem from allelic and epiallelic variation of genes and promoters. More recently, copy number variation (CNV) and specialisation in members of gene families have emerged as another important source of phenotypic variation. The application of bioinformatic tools has spurred the investigation of duplicate genes in the grapevine reference sequence. Genome-wide surveys of large gene families were performed for transcription factors of the MYB R2R3, MIKCC-type MADS box, AP2/ERF, and WOX homeodomain types (Matus et al. 2008; Díaz-Riquelme et al. 2009; Licausi et al. 2010; Gambino et al. 2011), aquaporins (Fouquet et al. 2008), Rab GTPases (Abbal et al. 2008), NB-LRR genes (Yang et al. 2008), PR-1 genes (Li et al. 2011), flavonoid hydroxylases (Falginella et al. 2010), histone acetyltransferases and deacetylases (Aquea et al. 2010), neutral invertases (Nonis et al. 2008), terpene synthases (Martin et al. 2010), stilbene synthases (Vannozzi et al. 2012), defensins (Giacomelli et al. 2012), smallRNAs (Carra et al. 2009), microRNAs (Pantaleo et al. 2010), ta-siRNAs (Zhang et al. 2012), SET DOMAIN GROUP (SDG) genes (Aquea et al. 2011), sugar transporters (Afoufa-Bastien et al. 2010) and

O-methyltransferases (Fournier-Level et al. 2011). Some studies also focused on structural genomics. To this end, the high quality of the PN40024 genome assembly helped to describe in unprecedented detail the molecular events that led to local expansion in copy number and divergence in the regulatory regions (Coleman et al. 2009; Falginella et al. 2010). Functional copy-number alteration is important in cancer predisposition and as a biomarker for tumour DNA. Given the similarity to plant somatic mosaicism, it is possible that the CNV of dosage-sensitive genes is another source of clonal variation in grapevine, and in particular with phenotypic effects in secondary metabolism.

8.11 Conclusions

New sequencing technologies offer a suite of tools in plant genomics to accelerate progress in the exploration of DNA variation. This will enable viticulturists and grapevine scientists to discern commonalities and diversities in the grape germplasm at the haplotype level and with higher resolution than ever. NGS-enabled technologies are now expected to parallel and complement past efforts of DNA genotyping for cataloguing of genetic variation and conservation of genetic resources.

8.12 References

Abbal P, Pradal M, Muniz L, Sauvage FX, Chatelet P, et al. (2008) Molecular characterization and expression analysis of the Rab GTPase family in *Vitis vinifera* reveal the specific expression of a VvRabA protein. *J Exp. Bot.* 59:2403–16.

Afoufa-Bastien D, Medici A, Jeauffre J, Coutos-Thévenot P, Lemoine R, et al. (2010) The *Vitis vinifera* sugar transporter gene family: phylogenetic overview and macroarray expression profiling. *BMC Plant Biol.* 10:245.

Aquea F, Timmermann T, Arce-Johnson P (2010) Analysis of histone acetyltransferase and deacetylase families of *Vitis vinifera. Plant Physiol. Biochem.* 48:194–9.

Aquea F, Vega A, Timmermann T, Poupin MJ, Arce-Johnson P (2011) Genome-wide analysis of the SET DOMAIN GROUP family in grapevine. *Plant Cell Rep.* 30:1087–97.

Aradhya MK, Dangl GS, Prins BH, Boursiquot JM, Walker MA, et al. (2003) Genetic structure and differentiation in cultivated grape, *Vitis vinifera* L. *Genet. Res.* 81:179–92.

Arrigo N, Arnold C (2007) Naturalised *Vitis* rootstocks in Europe and consequences to native wild grapevine. *PLoS ONE* 2(6):e521.

Arroyo-García R, Ruiz-García L, Bolling L, Ocete R, López MA, et al. (2006) Multiple origins of cultivated grapevine (*Vitis vinifera* L. ssp. *sativa*) based on chloroplast DNA polymorphisms. *Mol. Ecol.* 15:3707–14.

Benjak A, Boué S, Forneck A, Casacuberta JM (2009) Recent amplification and impact of MITEs on the genome of grapevine (*Vitis vinifera* L.). *Genome Biol. Evol.* 1:75–84.

Boss PK, Thomas MR (2002) Association of dwarfism and floral induction with a grape 'green revolution' mutation. *Nature* 416:847–50.

Boursiquot J-M, Lacombe T, Laucou V, Julliard S, Perrin F-X, et al. (2009) Parentage of Merlot and related winegrape cultivars of southwestern France: discovery of the missing link. *Aust. J. Grape Wine Res.* 15:144–55.

Bowers JE, Meredith CP (1997) The parentage of a classic wine grape, Cabernet Sauvignon. *Nat. Genet.* 16:84–7.

Bowers J, Boursiquot J-M, This P, Chu K, Johansson H, et al. (1999) Historical genetics: the parentage of Chardonnay, Gamay, and other wine grapes of Northeastern France. *Science* 285:1562–5.

Carra A, Mica E, Gambino G, Pindo M, Moser C, et al. (2009) Cloning and characterization of small non-coding RNAs from grape. *Plant J.* 59:750–63.

Carrier G, Le Cunff L, Dereeper A, Legrand D, Sabot F, et al. (2012) Transposable elements are a major cause of somatic polymorphism in. *Vitis vinifera* L. *PLoS One* 7:e32973.

Cipriani G, Spadotto A, Jurman I, Di Gaspero G, Crespan M, et al. (2010) The SSR-based molecular profile of 1005 grapevine (*Vitis vinifera* L.) accessions uncovers new synonymy and parentages, and reveals a large admixture amongst varieties of different geographic origin. *Theor. Appl. Genet.* 121:1569–85.

Coleman C, Copetti D, Cipriani G, Hoffmann S, Kozma P, et al. (2009) The powdery mildew resistance gene REN1 co-segregates with an NBS-LRR gene cluster in two Central Asian grapevines. *BMC Genetics* 10:89.

De Andrés MT, Benito A, Pérez-Rivera G, Ocete R, Lopez MA, et al. (2012) Genetic diversity of wild grapevine populations in Spain and their genetic relationships with cultivated grapevines. *Mol. Ecol.* 21:800–16.

Di Gaspero G, Cattonaro F (2010) Application of genomics to grapevine improvement. *Aust. J. Grape Wine Res.* 16:122–30.

Di Vecchi-Staraz M, Laucou V, Bruno G, Lacombe T, Gerber S, et al. (2009) Low level of pollen-mediated gene flow from cultivated to wild grapevine: consequences for the evolution of the endangered subspecies *Vitis vinifera* L. subsp. *sylvestris*. *J. Hered.* 100:66–75.

Falginella L, Castellarin SD, Testolin R, Gambetta GA, Morgante M, et al. (2010) Expansion and subfunctionalisation of flavonoid 3',5'-hydroxylases in the grapevine lineage. *BMC Genomics* 11:562.

Fernandez L, Torregrosa L, Segura V, Bouquet A, Martinez-Zapater JM (2010) Transposon-induced gene activation as a mechanism generating cluster shape somatic variation in grapevine. *Plant J.* 61:545–57.

Fouquet R, Léon C, Ollat N, Barrieu F (2008) Identification of grapevine aquaporins and expression analysis in developing berries. *Plant Cell Rep.* 27:1541–50.

Fournier-Level A, Lacombe T, Le Cunff L, Boursiquot JM, This P (2010) Evolution of the *VvMybA* gene family, the major determinant of berry colour in cultivated grapevine (*Vitis vinifera* L.). *Heredity* 104:351–62.

Fournier-Level A, Hugueney P, Verriès C, This P, Ageorges A (2011) Genetic mechanisms underlying the methylation level of anthocyanins in grape (*Vitis vinifera* L.). *BMC Plant Biol.* 11:179.

Gambino G, Minuto M, Boccacci P, Perrone I, Vallania R, et al. (2011) Characterization of expression dynamics of WOX homeodomain transcription factors during somatic embryogenesis in *Vitis vinifera*. *J. Exp. Bot.* 62:1089–101.

Giacomelli L, Nanni V, Lenzi L, Zhuang J, Dalla Serra M, et al. (2012) Identification and characterization of the defensin-like gene family of grapevine. *Mol. Plant-Microbe Interact.* 25:1118–31.

Hamblin MT, Buckler ES, Jannink JL (2011) Population genetics of genomics-based crop improvement methods. *Trends Genet.* 27:98–106.

Hocquigny S, Pelsy F, Dumas V, Kindt S, Héloir MC, et al. (2004) Diversification within grapevine cultivars goes through chimeric states. *Genome* 47:579–89.

Jaillon O, Aury JM, Noel B, Policriti A, Clepet C, et al. (2007) The grapevine genome sequence suggests ancestral hexaploidization in major angiosperm phyla. *Nature* 449:463–7.

Levadoux L (1956) Les populations sauvages et cultives de *Vitis vinifera* L. *Annales de l'Amélioration des Plantes* 1: 59–118.

Li ZT, Dhekney SA, Gray DJ (2011) PR-1 gene family of grapevine: a uniquely duplicated PR-1 gene from a *Vitis* interspecific hybrid confers high level resistance to bacterial disease in transgenic tobacco. *Plant Cell Rep.* 30:1–11.

Licausi F, Giorgi FM, Zenoni S, Osti F, Pezzotti M, et al. (2010) Genomic and transcriptomic analysis of the AP2/ERF superfamily in *Vitis vinifera*. *BMC Genomics* 11:719.

Mardis ER (2011) A decade's perspective on DNA sequencing technology. *Nature* 470:198–203.

Martin DM, Aubourg S, Schouwey MB, Daviet L, Schalk M, et al. (2010) Functional annotation, genome organization and phylogeny of the grapevine (*Vitis vinifera*) terpene synthase gene family based on genome assembly, FLcDNA cloning, and enzyme assays. *BMC Plant Biol.* 10:226.

Matus JT, Aquea F, Arce-Johnson P (2008) Analysis of the grape MYB R2R3 subfamily reveals expanded wine quality-related clades and conserved gene structure organization across *Vitis* and *Arabidopsis* genomes. *BMC Plant Biol.* 8:83.

Moncada X, Pelsy F, Merdinoglu D, Hinrichsen P (2006) Genetic diversity and geographical dispersal in grapevine clones revealed by microsatellite markers. *Genome* 49:1459–72.

Myles S, Chia JM, Hurwitz B, Simon C, Zhong GY, et al. (2010) Rapid genomic characterization of the genus *Vitis*. *PLoS One* 5(1):e8219.

Myles S, Boyko AR, Owens CL, Brown PJ, Grassi F, et al. (2011) Genetic structure and domestication history of the grape. *Proc. Natl. Acad. Sci. USA* 108:3530–5.

Nonis A, Ruperti B, Pierasco A, Canaguier A, Adam-Blondon AF, et al. (2008) Neutral invertases in grapevine and comparative analysis with *Arabidopsis*, poplar and rice. *Planta* 229:129–42.

Olmo HP (1996) The origin and domestication of the *V. vinifera* grape. In: McGovern P, Fleming SJ, Katz SH (eds) *The Origin and Ancient History of Wine*. Gordon and Breach Publishers, Amsterdam, the Netherlands. pp 31–43.

Pantaleo V, Szittya G, Moxon S, Miozzi L, Moulton V, et al. (2010) Identification of grapevine microRNAs and their targets using high-throughput sequencing and degradome analysis. *Plant J.* 62:960–76.

Pelsy F (2010) Molecular and cellular mechanisms of diversity within grapevine varieties. *Heredity* 104:331–40.

Péros J-P, Berger G, Portemont A, Boursiquot J-M, Lacombe T (2010) Genetic variation and biogeography of the disjunct *Vitis* subg. *Vitis* (Vitaceae). *J. Biogeography* 38:471–86.

Pindo M, Vezzulli S, Coppola G, Cartwright DA, Zharkikh A, et al. (2008) SNP high-throughput screening in grapevine using the SNPlex genotyping system. *BMC Plant Biol.* 8:12.

Pipia I, Gogniashvili M, Tabidze V, Beridze T, Gamkrelidze M, et al. (2012) Plastid DNA sequence diversity in wild grapevine samples (*Vitis vinifera* subsp. *sylvestris*) from the Caucasus region. *Vitis* 51:119–24.

This P, Lacombe T, Thomas MR (2006) Historical origins and genetic diversity of wine grapes. *Trends Genet.* 22:511–09.

Vannozzi A, Dry IB, Fasoli M, Zenoni S, Lucchin M (2012) Genome-wide analysis of the grapevine stilbene synthase multigenic family: genomic organization and expression profiles upon biotic and abiotic stresses. *BMC Plant Biol.* 12(1):130.

Velasco R, Zharkikh A, Troggio M, Cartwright DA, Cestaro A, et al. (2007) A high quality draft consensus sequence of the genome of a heterozygous grapevine variety. *PLoS One* 2:e1326.

Vezzulli S, Micheletti D, Riaz S, Pindo M, Viola R, et al. (2008) A SNP transferability survey within the genus *Vitis*. *BMC Plant Biol* 8:128.

Williams LJS, Tabbaa DG, Li N, Berlin AM, Shea TP, et al. (2012) Paired-end sequencing of Fosmid libraries by Illumina. *Genome Res.* 22(11):2241–9.

Yakushiji H, Kobayashi S, Goto-Yamamoto N, Tae Jeong S, Sueta T, et al. (2006) A skin colour mutation of grapevine, from black-skinned Pinot Noir to white-skinned Pinot Blanc, is caused by deletion of the functional VvmybA1 allele. *Biosci. Biotechnol. Biochem.* 70:1506–8.

Yang S, Zhang X, Yue JX, Tian D, Chen JQ (2008) Recent duplications dominate NBS-encoding gene expansion in two woody species. *Mol. Genet. Genomics* 280:187–98.

Zhang C, Li G, Wang J, Fang J (2012) Identification of trans-acting siRNAs and their regulatory cascades in grapevine. *Bioinformatics* 28(20):2561–8.

Zohary D (1996) The domestication of *Vitis vinifera* L. in the Near East. In: McGovern P, Fleming SJ, Katz SH (eds) *The Origin and Ancient History of Wine*. Gordon and Breach Publishers, Amsterdam, the Netherlands. pp. 21–8.

9

Peach ripening transcriptomics unveils new and unexpected targets for the improvement of drupe quality

Nicola Busatto, Md Abdur Rahim and Livio Trainotti, University of Padova, Italy

DOI: 10.1533/9781908818478.165

Abstract: Peach is a climacteric fruit, and ethylene plays a critical role during ripening. The dramatic increase in ethylene production that leads to drupe ripening and softening hardly affects the possibility of long storage periods. Several post-harvest diseases affect peach fruit that has undergone post-harvest treatments such as cold, thus limiting its fresh consumption during the production season and its distant shipping (mainly from the southern to the northern hemisphere). Moreover, the use of chemical compounds, mostly the ethylene receptor inhibitor 1-methylcyclopropene (1-MCP), widely used to control ripening in several climacteric fruit, is ineffective with peach. Extensive transcriptome profiling has evidenced the uniqueness of the activation of climacteric ethylene production in peach and other drupes, such as apricot. The tight synchronization between the activation of the auxin and ethylene pathways, the cross-talk between the two and the existence of genes uniquely modulated by one of the two hormones highlight new possible targets for genetic improvement and new possibilities to chemically regulate peach ripening.

Key words: peach ripening, microarrays, transcriptome, root growth factor, peptide, hormones.

Abbreviations: ACC, 1-aminocyclopropane-1-carboxylic acid; 1-MCP, 1-methylcyclopropene; NAA, 1-naphthalene acetic acid;

NCED, 9-cis-epoxycarotenoid dioxygenase; ABA, abscisic acid; ACO, ACC oxidase; ACS, ACC synthase; ARE, Auxin Responsive Element; EXP, expansin; EST, expressed sequence tag; IAA, indole acetic acid; LRR-RLK, leucine-rich repeat receptor-like kinase; PME, pectin methylesterase; PG, polygalacturonase; RNAseq, RNA sequencing; RGL, root growth factor-like; RGF, root growth factor.

9.1 Introduction

The peach (*Prunus persica* L. (Batsch)) tree produces fruits that are appreciated worldwide, either consumed fresh or used by industry to prepare canned products, jams and juices. Global peach and nectarine production is more than 20 Mtonnes/year (Food and Agriculture Organization Statistics (FAOSTAT) 2012 data), half of it concentrated in China. Italy, Spain and the USA follow, each producing more than 1 Mtonne/year. Many other stone fruit species (most, like peach, domesticated in China and belonging to the *Prunus* genus) are widely cultivated worldwide, but peach is by far the most important. It has a small genome (eight chromosomes for an estimated nuclear DNA content recently assessed at ~230 Mbp; Sosinski et al., 2010) and biological and genetic characteristics, such as self-compatibility and a relatively short juvenile period, that made this species a model for stone fruits. Genetic studies date back to the 1980s, with the first maps published in the 1990s (reviewed in Verde et al., 2012). In 2010 the first draft of the genome assembly (Peach V1.0) was released (Sosinski et al., 2010). A total of 202 scaffolds larger than 1Kb, equaling approximately 227 Mbp, a bit less than the 280–300 Mbp that was previously predicted, were aligned to eight pseudomolecules, named according to their corresponding linkage groups and comprising 99.3% of the peach genome, using a network of *Prunus* linkage maps. Gene prediction and annotation algorithms produced 27 851 genes, 838 of which show alternative splicing, thus giving a total of 28 689 predicted protein-coding genes.

Before reaching the milestone of the peach genome sequence, several transcriptomics approaches have been undertaken in the *Prunus* genus since the beginning of the new century. In a few years, a set of tools have been developed and used, mainly in peach, apricot and almond. Transcriptomics tools have been primarily used to investigate fruit ripening and post-harvest physiology, but also disease resistance and flower transition. At the beginning of the second decade of the century, more than 100 000 ESTs are available in public databases, the majority of which were obtained from peach fruit. This repertoire has been used for digital expression analyses of transcriptome changes associated with fruit ripening and the appearance of chilling-induced post-harvest fruit disorders. The microarrays developed on peach expressed sequence tags (ESTs) have been extensively used, mainly to investigate fruit biology, but are soon going to be replaced by genome-wide platforms based on the recently released peach genome sequence (Trainotti et al., 2012). In this chapter the main outcomes of several transcriptomics approaches undertaken to shed light on peach fruit ripening will be summarized, and possible traditional and biotechnological methodologies to improve drupe quality will be presented and discussed.

9.2 The fruit

Terrestrial vascular plants are sessile organisms, and therefore they accomplish their entire life in the same place where they germinate and develop. In particular, species belonging to the *Spermatophyta* division have introduced an essential innovation: the seed. This represents a typical diffusion means for higher plants, and it is the result of a reproductive act. The seed could be interpreted as a resistance apparatus whereby a young individual, protected and nourished by different structures, is separated from the mother plant. To make the most of the energetic effort made during seed production, it is necessary that seed dispersal is as efficient as possible to avoid competition for resources such as light, water or nutrients in the soil.

To answer this necessity, the Spermatophytes evolved different strategies and the Angiosperms adopted the fruit as solution, with a wide range of heterogeneous shapes related to the different dispersal methods used. For instance, the winged dry indehiscent samaras of maples are developed expressly to be dispersed by wind. Another dry fruit, the dehiscent *Arabidopsis* silique, releases the seeds by opening valves that themselves remain with the silique.

A totally different approach is represented by fleshy fruits that are dispersed via ingestion by vertebrate animals. This strategy is named endozoochory, and it is generally the result of a mutualistic co-evolution in which a plant surrounds seeds with an edible, nutritious tissue as a good food for animals that consume it, carrying the seeds far away from the mother plant.

The transition from an immature and unattractive fruit to a desirable and edible food is caused by the ripening process, which consists of biochemical and physiological changes. These modifications, although variable among species, generally include conversion of starch to sugars, modification of cell wall structure and texture leading to pulp softening, alterations in pigmentation, and accumulation of flavor and aromatic volatiles (Giovannoni, 2001).

Among fleshy fruits two different physiological categories, climacteric and non-climacteric, can be found. Climacteric fruits present an increase in the respiration rate before the visible onset of the ripening process, accompanied by a spike in the production of ethylene, which is recognized as the hormone that accelerates the ripening of fruits. In non-climacteric fruits the respiration and ethylene peaks are not detectable and exogenous applications of the gaseous hormone do not trigger the process. Different examples of climacteric fruits are tomatoes, apples, pears, apricots, kiwis and peaches, while strawberries, oranges, olives, watermelon and grapes belong to the non-climacteric group.

Ripening of fleshy fruits is a complex and highly coordinated developmental process that usually, but not always, coincides with seed maturation. The modifications that involve the fruit during the ripening process lead to the modulation of the expression levels of thousands of genes (Alba et al., 2005). As previously mentioned, all these changes are related to the fruit's transition from an immature stage to a succulent and attractive food for the animals that consume it.

For this purpose the ripening process leads to the following changes:

a) the fruit pigmentation operated by storage of anthocyanic vacuolar inclusions;

b) the chloroplast turning into chromoplast, with accumulation of pigments such as lycopene or β-carotene and the associated degradation of chlorophyll;

c) the conversion of starch accumulated during fruit development to glucose and fructose, the main sugars present in ripe fruit, which can represent up to 4% of the fresh weight of the fruit;

d) abundant synthesis of organic acids such as malic and citric acids, which, with sugars, give a good flavor;

e) synthesis of volatile compounds;

f) softening, which involves texture modifications due to the action of several cell wall enzymes, which mirror the complexity of cell wall composition: polygalacturonase (PG), pectin methylesterase (PME) and expansin (EXP), to give a few examples;

g) a generally enhanced susceptibility to opportunistic pathogens, likely associated with the loss of cell wall integrity.

9.3 Peach development and ripening

Peaches are climacteric fruits. The period of time for the development of *Prunus persica* fruits can be divided into four stages, from S1 to S4 (Masia et al., 1992). The double-sigmoid curve is characteristic of drupe development, in which stages of fast growth (S1 and S3) are followed by stages in which growth is slow (S2 and S4). During each stage major cellular, biochemical and physiological processes occur. In particular, in S1 both cellular multiplication and distension occur; during S2 the hardening of the endocarp (pith) takes place and, because of this, growth slows down to recover during S3, when the fruit increase is mainly due to cell expansion. Finally, in S4 the fruit reaches its final dimensions and ripens.

Peaches exhibit a sharp rise in ethylene production at the onset of ripening, paralleled by dramatic changes in the transcriptional profile of genes, many of them regulated by the hormone (Trainotti et al., 2007). An important role in the coordination of mesocarp development and ripening is played by the seed (Bonghi et al., 2011). Such coordinated and programmed modulation of gene expression leads to several changes, all of which contribute to overall fruit quality (Trainotti et al., 2003, 2006). In peach fruit, there is a close link between on-tree physiological maturity and development of key traits responsible for its quality. A delayed harvest could improve fruit organoleptic characteristics, since sugars and flavor components increase while total acids decrease during late ripening (Vizzotto et al., 1996; Visai and Vanoli, 1997; Boudehri et al., 2009). However, melting flesh peaches and nectarines undergo rapid ripening and soften quickly after harvest, leading to losses in the market chain. Therefore, fruits are commonly picked at an early stage of ripening to better withstand handling, and they do not reach full flavor and aroma. Furthermore, chilling injury (CI) and diseases arising in peach fruit that have undergone post-harvest cold storage are among the main problems for the peach industry; thus keeping fruit at low temperature is a solution that cannot always or easily be practiced for this commodity (Vizoso et al., 2009).

The ethylene receptor inhibitor 1-methylcyclopropene (1-MCP) is widely used for the storage of climacteric fruit as, by blocking the hormone's action, it prevents the deleterious effects of over-ripening (Watkins, 2006). This molecule has also been widely tested to improve storage of peaches and nectarines, but its effects are conflicting (Blankenship and Dole, 2003). Indeed, 1-MCP delays the loss of flesh firmness but does not block ethylene biosynthesis, which, on the contrary, is stimulated. Thus the final effect of the chemical application is fruit producing more ethylene (Ziliotto et al., 2008), and so undergoing rapid rotting. This unexpected effect of 1-MCP aroused the curiosity of peach researchers, who applied transcriptomics tools to try to unveil the molecular basis of the phenomenon (Ziliotto et al., 2008; Tadiello, 2010), thus coming across a potentially new class of hormone peptides, whose manipulation might offer new possibilities to improve peach fruit quality and storage (Busatto, 2012).

9.4 Microarray Transcript Profiling in peach

The beginning of the modern transcriptomics era in *Prunus* spp. might be traced back to the first report describing EST production from peach, dating back to the year 2000 (Hayama et al., 2000), when about 200 sequences were obtained from libraries constructed from RNA of three different fruit developmental stages. Soon after, several laboratories developed EST collections that were used to produce nylon macroarrays (Trainotti et al., 2003; González-Agüero et al., 2008) and, later on, glass microarrays (ESTree Consortium, 2005). Both array (Trainotti et al., 2003, 2006, 2007; Ogundiwin et al., 2008), digital gene expression profiling (Vecchietti et al., 2008; Vizoso et al., 2009) and large-scale qRT-PCR approaches (Pegoraro et al., 2010) were used to look for the genetic determinants of fruit quality and genes associated with post-harvest diseases. These approaches highlighted that transcripts more abundant in the skin contained sequences coding for lipid transfer proteins, the major protein allergen (which, besides being the most abundant in ripe fruit, seems to be specific to the genus), and a 9-cis-epoxycarotenoid dioxygenase (NCED), a key enzyme in abscisic acid (ABA) synthesis. Furthermore, transcripts encoding proteins involved in synthesis of volatiles were more frequently cloned in the skin, confirming that the peel is central in peach flavor production (Vecchietti et al., 2008).

With the aim of identifying genes related to disorders caused by cold storage, more than 41 000 ESTs have been obtained from four different libraries highlighting 197 transcripts that significantly differed in abundance in at least two conditions (Vizoso et al., 2009). The analysis revealed that ripening has a stronger effect on the peach transcriptome than cold treatment, and highlighted the relevance of genes associated with oxidative stress and metabolic imbalances occurring mainly in the organelles (mitochondria and plastids), but also at the level of the plasma membrane. These findings opened new fields for future research because they added new candidates to the list of genes/metabolic processes causing post-harvest diseases.

µPEACH1.0 (ESTree Consortium, 2005) and ChillPeach (Ogundiwin et al., 2008) are the two main microarray platforms that have been developed from *Prunus* derived ESTs. µPEACH1.0 consists of 4806 80-mer oligonucleotide probes. The transcript set used to design the probes was biased towards genes more actively transcribed in the mesocarp. Nevertheless, since ripening and fruit storage were the main interests of the researchers using this tool, it allowed relevant findings in these fields. A dramatic change in the mesocarp transcriptome occurs during the S3 to S4 transition, with hundreds of genes that modulate their expression (Trainotti et al., 2006, 2007). Among the differentially transcribed genes, many belong to the cell wall and chloroplast compartments. Auxin and ethylene treatments of pre-climacteric fruits allowed a new and independent role during ripening of this climacteric fruit to be assigned to auxin, with several *Aux/IAA* genes involved in the process and differently regulated by the two hormones (Trainotti et al., 2007). The role of ethylene and auxin during ripening has also been confirmed by transcriptome profiling of fruit treated with the ethylene inhibitor 1-MCP (Ziliotto et al., 2008; Tadiello, 2010). This intense auxin–ethylene cross-talk during mesocarp ripening does not seem to be exclusive to peach, since the up-regulation of genes belonging to the auxin domain has also been documented in apricot, by both in silico northern (Grimplet et al., 2005) and microarray profiling (Manganaris et al., 2011). The effect of jasmonates (JAs), applied to fruit on the tree at the S3 stage, has also been investigated using µPEACH1.0 (Ziosi et al., 2008). Transcriptome profiling revealed a profound effect of JAs, which repressed ethylene and auxin-related genes and, as a consequence, also ripening, while inducing stress and defense-related genes, keeping fruit cells in a more metabolically active state.

The availability of the peach genome poses new challenges to transcriptomics in this and in closely related species. New technologies are being applied, and some, such as RNA sequencing (RNAseq) with different platforms, have already been successfully used in peach, almond and apricot (Martínez-Gómez et al., 2011). Microarray hybridizations will still be used, as they can compete for data quality and prices with the new DNA/RNA sequencing technologies.

A new microarray platform, named µPEACH3.0, has recently been developed and will be used in the near future. It is based on a custom-made Agilent SurePrint G3 platform with a 8 × 60 k format. The probe selection has been carried out on 30 113 predicted transcripts. The 28 689 officially released transcripts (Sosinski et al., 2010) have been implemented by adding 1424 new predictions (Forcato et al., unpublished data). By using this new platform, a new precious expression dataset was made available, obtained by several microarray experiments performed with the goal of identifying genes involved in seed–pericarp cross-talk and mesocarp development in peach (Bonghi et al., unpublished data). This allowed a complete and detailed kinetics of expression profiles to be obtained for a massive number of genes during peach development and ripening. Expression profiles of several genes (*ACSs, ACOs, LRR-RLKs*; Figure 9.1) shown here have been extrapolated from microarray data obtained with this platform.

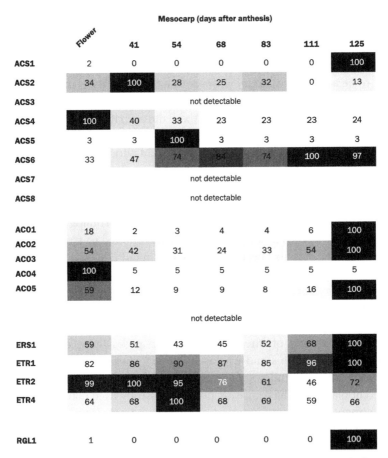

Figure 9.1 Expression profiles of ethylene-related genes during fruit development and ripening. Expression profiles were measured in flower and mesocarp at six different developmental stages (S1 to S4). Expression values are related to the highest expression of each gene (100% black)

Ethylene synthesis during the fruit climacteric has been largely investigated in peach, and several *ACS* (Tatsuki et al., 2006) and *ACO* genes (Ruperti et al., 2001) have been characterized and cloned. Peach seems to be a peculiar fruit, with the only *ACS1* involved in both system1 and system2 ethylene biosynthesis, while in other species such as tomato (Barry and Giovannoni, 2007) and apple (Wang et al., 2009) more *ACS* genes contribute differently to the tight regulation of hormone synthesis.

With the appearance of the genome sequence, five more peach *ACS* genes have been identified (Figure 9.2). Nonetheless, initial expression analyses confirmed that *ACS1* is the gene that is most likely responsible in the fruit for ethylene production during both vegetative growth

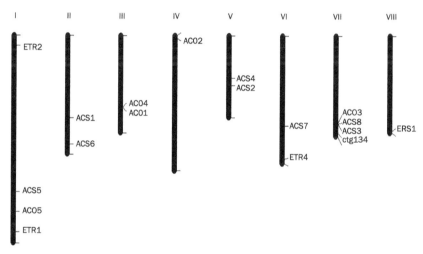

Figure 9.2 Chromosomal location of ethylene-related genes on the peach physical map. The map and the gene nomenclature are according to Sosinski et al. (2010). Gene names and identification codes are: ACS1, ppa004774m; ACS2, ppa016458m; ACS3, ppa008124m; ACS4, ppa003908m; ACS5,ppa015636m; ACS7, ppa004987m; ACS8, ppa022214m; ACS6, ppa004475m; ACO1, ppa008791m; ACO2, ppa008813m; ACO4, ppa022135m; ACO3, ppa009228m; ACO5, ppa010361m; ETR1, ppa002851m; ETR1, ppa001917m; ETR4, ppa001846m; ERS1, ppa002692m; ETR2, ppa001786m; CTG134 (or RGL1), ppa012311m

(system1) and climacteric (system2). Indeed, beside *ACS1*, only *ACS4* (Figure 9.1), out of the eight genes so far identified, is expressed at detectable levels during peach fruit development and ripening. But it is questionable whether *ACS4* codes for an authentic ACS rather than a simple aminotransferase, as it is closely related to *Arabidopsis ACS10* and *ACS12*, which have been shown to lack ACS activity (Yamagami et al., 2003).

Thus, if *ACS4* cannot be considered, *ACS2, ACS5* and *ACS6* are left. But their expression is very low (at maximum about 3, 0.5 and 1% of *ACS1*, respectively) and thus their contribution to ethylene synthesis might be questionable. It is thus possible to speculate that the tight regulation of *ACS1* expression alone might be responsible for the whole ethylene production, at least in the fruit. More detailed biochemical and genetic studies might contribute to solving this question.

9.5 New players in the control of peach ripening

In multicellular organisms, cell-to-cell communication is essential to regulate and organize complex processes such as reproduction or growth. In addition to traditional hormones, a new class of secretory and non-secretory peptides, involved in several aspects of plant growth regulation, including defense responses, callus growth, meristem organization, self-incompatibility (SI), root growth, leaf-shape regulation, nodule development and organ abscission, has emerged (Yamagami et al., 2003).

Tomato Systemin (SYS) was the first to be discovered, 20 years ago; it is involved in the response against insect attacks by induction of the systemic defense (Pearce et al., 1991). Peptide hormones can also regulate cell proliferation, as in the case of phytosulfokine (PSK), a five amino acid peptide that acts as a mitogenic factor (Igasaki et al., 2003) promoting tracheary element (TE) differentiation of dispersed *Zinnia* mesophyll cells, adventitious bud formation, adventitious root formation, and pollen germination. Another well-known peptide hormone is Clavata 3 (Clark et al., 1995), which is necessary in *Arabidopsis* to maintain the correct shape and function of the shoot apical meristem. In particular, *CLV3* encodes a 96 amino acid peptide with a secretion signal at the N terminus. In general, these genes encode peptides whose sequence presents a high level of diversity, with the exception of the conserved domains that correspond to the mature peptide sequences. The synthesis pathway of secreted peptide hormones begins with the translation of a pre-propeptide, followed by post-translational modifications, the cleavage of the N-terminal signal peptide and the proteolytic processing leading to the active hormone.

Any hormone must have a receptor, and so far the identified peptide hormone receptors belong to the wide multi-gene family (more than 600 members in *Arabidopsis*) of receptor like kinases (RLKs; Shiu and Bleecker, 2003). These proteins have an extracellular N-terminal domain that includes the signal peptide, a ligand-binding domain and a transmembrane portion. The cytoplasmic C-terminal domain acts as a serine threonine kinase to initiate the signal transduction cascade. The RLKs have been categorized into four main families: S-domain class, EGF-like, LecRK (Lectin Receptor Kinase) and LRR-RLK (Leucine-Rich-Repeat), which comprises the largest class of plant RLKs (235 members in *Arabidopsis*). In particular, receptors of some peptide hormones, such as SYS, CLV3, PSK, IDA (inflorescence deficient in abscission) (Matsubayashi and Sakagami, 2006) and PSY1 (Amano et al., 2007), belong to the LRR-RLK class.

The peach CTG134 was identified by means of microarray experiments that were carried out during studies focused on peach ripening (Trainotti et al., 2007; Tadiello, 2010). It was noticed because it was the only gene to be induced by ripening, auxin and 1-MCP, while most of the genes associated with ripening are repressed by the ethylene inhibitor. Moreover, this microarray result was confirmed by real time PCR data, demonstrating that CTG134 was specifically expressed at the S3–S4 transition (Figure 9.1; see RGL1), was up-regulated by NAA (1-naphthalene acetic acid, an auxin analog), repressed by ethylene in pre-climacteric S3 fruit and stimulated by 1-MCP.

The corresponding cDNA was isolated and sequenced. It encodes a 174 amino acid protein with a predicted molecular mass of 18.5 kDa. The prediction indicates the presence of a highly hydrophobic signal peptide at the N-terminal domain that drives CTG134 protein into the apoplast. The CTG134 amino acid sequence shows no similarity to other known proteins except for its C-terminal domain, which is very well conserved if compared with those of root growth factor (RGF) peptide hormones (Matsuzaki et al., 2010). It shares an Asp-Tyr motif essential for post-translational sulfation. Although peptide hormones are known to be involved in a wide range of biological processes, none has previously been associated with the fruit ripening process, as is the case with the peach CTG134.

The CTG134 auxin responsiveness has been confirmed by promoter::GUS fusion experiments. The *CTG134* promoter sensitivity to auxin and the two fully functional Auxin Responsive Elements (ARE) present on it were confirmed by means of the beta-glucuronidase (GUS) reporter gene in heterologous systems such as *Nicotiana tabacum* and *Arabidopsis thaliana*. A crucial finding on the promoter of *CTG134* is its positive response to both auxin and 1-MCP (Busatto, 2012). While the first can be easily explained by the presence of several ARE, the latter is more intriguing and may have physiological relevance only if the altered perception of ethylene can either de-repress *CTG134* transcription or induce the concentration of a potential activator such as auxin. It is noteworthy that in peach mesocarp treated with 1-MCP not only the induction of *CTG134* was observed, but also that of several auxin responsive genes, such as a *GH3* and several genes encoding Aux/IAA proteins (Tadiello, 2010).

The analysis of tobacco transgenic lines overexpressing *CTG134* allowed various phenotypes to be observed, in particular in capsules and roots. The roots of 35S::CTG134 plants displayed enhanced growth of root hairs, whose number is increased due to the transformation of atrichoblasts into trichoblasts and which are on average three times longer than in controls. In addition, an induction of root primordia formation but an inhibition of early root elongation was observed. Moreover, capsules, before the final drying, are 16% bigger than the controls.

The phenotypes observed in the *35S::CTG134* plants are all related to the ethylene domain and can be explained either by an increased synthesis of endogenous hormone or by an increased sensitivity to ethylene. Molecular analyses on *ACO* gene expression profiles seem to exclude an effect on ethylene synthesis, inducing speculation about an enhanced sensitivity to the gaseous hormone. When *CTG134* is overexpressed in *Arabidopsis* or tobacco, or if the mature peptide is exogenously added to the growth medium, it induces the formation of supernumerary root hairs. Furthermore, it alters the gravitropic response (Busatto, 2012), as recently shown for the GOLVEN peptides (Whitford et al., 2012), another name given to RGFs. Based on literature data (Matsuzaki et al., 2010; Whitford et al., 2012) and research findings (Busatto, 2012), we speculate that CTG134 is a peptide hormone and could be a mediator in the auxin/ethylene interplay occurring at the onset of peach ripening. For this reason CTG134 has been named RGL1, the first peach Root Growth factor-Like peptide discovered up to now.

The data gained from plants overexpressing RGL1 and with the external application of mature peptide suggest that RGL1 enhances the sensitivity to ethylene rather than its biosynthesis. Moreover, the ethylene treatment of peach fruit represses RGL1 expression (Tadiello, 2010) and thus it is more likely that the protein is synthesized when the ethylene level is low, allowing the plant to better sense the gaseous hormone just before the climacteric phase. Thereafter, when RGL1 action is no longer necessary as ethylene becomes abundant, *RGL1* transcription ceases and the protein is no longer produced. On the other hand, at the pre-climacteric stage, the occupancy of the ethylene receptor(s) by an inhibitor (1-MCP) enhances *RGL1* transcription. In tomato the level of ethylene receptors is important in determining the onset of fruit ripening, as demonstrated by the early-ripening phenotype of fruits produced by plants in which *LeETR4* or *LeETR6* are silenced (Kevany et al., 2007). In peach, too, the expression of the ethylene receptors in fruits is tightly regulated (Figure 9.1) (Rasori et al., 2002; Trainotti et al., 2007), with *ETR2* and *ERS1* increasing most during ripening and being slightly repressed by 1-MCP (Tadiello, 2010). It is not known whether the RGL1 peptide decreases ethylene receptor levels, thus leading to an increased hormone sensitivity, or causes a release of free auxin, leading to the induction of auxin-responsive genes such as *ACS1, RGL1* and several *Aux/IAA* genes (Tadiello, 2010), but the induction of *RGL1* transcription by 1-MCP links the two pathways.

Just as root needs to reduce the perception of ethylene induced by the presence of auxin and cytokines (Chilley et al., 2006), it is possible to suppose that a climacteric fleshy fruit such as peach must have the opposite need. Indeed, a well-modulated ethylene perception allows a homogeneous response into the whole pericarp during ripening and a correct transition from system 1 to system 2 ethylene synthesis. This suggests a role for the peptide in modulating this transition, also considering its auxin responsiveness together with IAA induction of *ACS1*, the key gene in climacteric ethylene production (Trainotti et al., 2007). However, the molecular and biochemical action of RGL1 remains obscure.

A possible way to get more information on this aspect might be to identify a putative receptor for RGL1, with the aim of investigating its effect on ethylene perception and the hormone signal transduction. As described above, all the known receptors for small post-translationally modified peptide signals belong to the LRR-LRK family, such as those for SYS, CLV3 or PSK (Matsubayashi and Sakagami, 2006; Amano et al., 2007). For this reason it is probable that the RGL1 receptor might also be a member of this group, considering the similarities that RGL1 has in common with other sulfated peptide hormones.

The recently produced expression dataset, obtained by several microarray experiments performed with the aim of identifying genes involved in seed–pericarp cross-talk and mesocarp development in peach (Bonghi et al., unpublished data), allowed a complete and detailed kinetics of expression profiles to be obtained for a massive number of genes during peach development and ripening. The mining of these data provided insight into the identification of a putative receptor for RGL1. Crossing these data with those used for the annotation of the peach genome, which provides a significant link between *Arabidopsis* and peach genes, it was possible to identify 391 LRR-RLK genes spotted on the array. Some of these are annotated as PSK and PSY receptors, and, interestingly, they are up-regulated during the ripening process. Further investigations, such as validations by Real-Time PCR, will be necessary to confirm these expression profiles and to select from the candidates the most likely receptor for RGL1.

9.6 Conclusions

Climacteric fruit needs a large amount of ethylene during the ripening process, and in some species, such as peach, its biosynthesis is stimulated by a rise in auxin slightly preceding the climacteric peak. For this reason, a positively modulated perception of ethylene might represent an advantage for the pre-climacteric stage, allowing the switch from system 1 to system 2 synthesis that leads to the climacteric phase and thus ripening. The finding that *RGL1* is not expressed in peach cultivars whose fruits do not produce climacteric ethylene (Tadiello and Trainotti, unpublished results) makes it a good target for developing new ways to increase and better control fruit quality. Genetic variation at loci controlling this pathway might be scanned to look for those allelic variants that might impact fruit ripening and quality, thus allowing development of cultivars that better withstand ripening and storage. Genomics tools such as SNP profiling platform and expression arrays are already available for peach and will have a great impact in the coming years. Moreover, the development of treatments based either on synthetic peptides or on other classes of molecules that interfere with the hormone/receptor binding and/or activities could also allow control of the timing of fruit ripening and, ultimately, its quality. This field is still in its infancy, and more basic research is needed before it can lead to practical applications, but, interestingly, synthetic peptides are relatively cheap to produce on an industrial scale, and thus the development of new agrochemicals based on this class of compounds is a very promising sector for modern agriculture.

9.7 Acknowledgements

Research in LT's laboratory is supported by Italian MIUR and University of Padua. MAR is supported by a 'Fondazione CARIPARO' fellowship.

9.8 References

Alba, R., Payton, P., Fei, Z., McQuinn, R., Debbie, P. et al. (2005). Transcriptome and selected metabolite analyses reveal multiple points of ethylene control during tomato fruit development. *Plant Cell* 17(11): 2954–2965.

Amano, Y., Tsubouchi, H., Shinohara, H., Ogawa, M. and Matsubayashi, Y. (2007). Tyrosine-sulfated glycopeptide involved in cellular proliferation and expansion in Arabidopsis. *Proceedings of the National Academy of Sciences of the USA* 104: 18333–18338.

Barry, C.S. and Giovannoni, J.J. (2007). Ethylene and fruit ripening. *Journal of Plant Growth Regulation* 26: 143–59.

Blankenship, S.M. and Dole, J.M. (2003). 1-Methylcyclopropene: a review. *Postharvest Biology and Technology* 28: 1–25.

Bonghi, C., Trainotti, L., Botton, A., Tadiello, A., Rasori, A. et al. (2011). A microarray approach to identify genes involved in seed-pericarp cross-talk and development in peach. *BMC Plant Biology* 11: 107.

Boudehri, K., Bendahmane, A., Cardinet, G., Troadec, C., Moing, A. and Dirlewanger, E. (2009). Phenotypic and fine genetic characterization of the D locus controlling fruit acidity in peach. *BMC Plant Biology* 9: 59.

Busatto, N. (2012). Functional characterization of a ripening induced RGF-like peptide hormone in peach. PhD thesis, University of Padua, Italy.

Chilley, P.M., Casson, S.A., Tarkowski, P., Hawkins, N., Wang, K.L. et al. (2006). The POLARIS peptide of Arabidopsis regulates auxin transport and root growth via effects on ethylene signaling. *Plant Cell* 18: 3058–72.

Clark, S.E., Running, M.P. and Meyerowitz, E.M. (1995). CLAVATA3 is a specific regulator of shoot and floral meristem development affecting the same processes as CLAVATA1. *Development* 121: 2057–67.

ESTree Consortium (2005). Development of an oligo-based microarray (µPEACH 1.0) for genomics studies in peach fruit. *Acta Horticulturae* 682: 263–8.

Giovannoni, J.J. (2001). Molecular regulation of fruit ripening. *Annual Review of Plant Physiology and Plant Molecular Biology* 52: 725–49.

González-Agüero, M., Pavez, F., Ibáñez, L., Pacheco, I., Campos-Vargas, R. et al. (2008). Identification of woolliness response genes in peach fruit after post-harvest treatments. *Journal of Experimental Botany* 59: 1973–86.

Grimplet, J., Gaspar, J.W., Gancel, A.L., Sauvage, F.X. and Romieu, C. (2005). Including mutations from conceptually translated expressed sequence tags into orthologous proteins improves the preliminary assignment of peptide mass fingerprints on non-model genomes. *Proteomics* 5: 2769–77.

Hayama, H., Shimada, T., Yamamoto, T., Iketani, H., Matsuta, N. et al. (2000). Characterization of randomly obtained cDNAs from peach fruits at various developmental stages. *Journal of the Japanese Society for Horticultural Science* 69: 183–5.

Igasaki, T., Akashi, N., Ujino-Ihara, T., Matsubayashi, Y., Sakagami, Y. and Shinohara, K. (2003). Phytosulfokine stimulates somatic embryogenesis in Cryptomeria japonica. *Plant Cell Physiology* 44: 1412–16.

Kevany, B.M., Tieman, D.M., Taylor, M.G., Dal Cin, V. and Klee, H.J. (2007). Ethylene receptor degradation controls the timing of ripening in tomato fruit. *The Plant Journal* 51: 458–67.

Manganaris, G.A., Rasori, A., Bassi, D., Geuna, F., Ramina, A. et al. (2011). Comparative transcript profiling of apricot (Prunus armeniaca L.) fruit development and on-tree ripening. *Tree Genetics & Genomes* 7: 609–16.

Martínez-Gómez, P., Crisosto, C.H., Bonghi, C. and Rubio, M. (2011). New approaches to Prunus transcriptome analysis. *Genetica* 139: 755–69.

Masia, A., Zanchin, A., Rascio, N. and Ramina, A. (1992). Some biochemical and ultrastructural aspects of peach fruit development. *Journal of the American Society for Horticultural Science* 117: 808–15.

Matsubayashi, Y. and Sakagami, Y. (2006). Peptide hormones in plants. *Annual Review of Plant Biology* 57: 649–74.

Matsuzaki, Y., Ogawa-Ohnishi, M., Mori, A. and Matsubayashi, Y. (2010). Secreted peptide signals required for maintenance of root stem cell niche in Arabidopsis. *Science* 329: 1065–7.

Ogundiwin, E.A., Martí, C., Forment, J., Pons, C., Granell, A. et al. (2008). Development of ChillPeach genomic tools and identification of cold-responsive genes in peach fruit. *Plant Molecular Biology* 68: 379–97.

Pearce, G., Strydom, D., Johnson, S. and Ryan, C.A. (1991). A polypeptide from tomato leaves induces wound-inducible proteinase inhibitor proteins. *Science* 253: 895–7.

Pegoraro, C., Roggia Zanuzo, M., Chaves, C.F., Brackmann, A., Girardi, C.L. et al. (2010). Physiological and molecular changes associated with prevention of woolliness in peach following pre-harvest application of gibberellic acid. *Postharvest Biology Technology* 57: 19–26.

Rasori, A., Ruperti, B., Bonghi, C., Tonutti, P. and Ramina, A. (2002). Characterization of two putative ethylene receptor genes expressed during peach fruit development and abscission. *Journal of Experimental Botany* 53: 2333–9.

Ruperti, B., Bonghi, C., Rasori, A., Ramina, A. and Tonutti, P. (2001). Characterization and expression of two members of the peach 1-aminocyclopropane-1-carboxylate oxidase gene family. *Physiologia Plantarum* 111: 336–44.

Shiu, S.-H. and Bleecker, A.B. (2003). Expansion of the receptor-like kinase/pelle gene family and receptor-like proteins in Arabidopsis. *Plant Physiology* 132: 530–43.

Sosinski, B., Verde, I., Morgante, M. and Rokhsar, D. (2010). The international peach genome initiative. A first draft of the peach genome sequence and its use for genetic diversity analysis in peach. In: *5th International Rosaceae genomics conference* (Stellenbosch, South Africa), page O46.

Tadiello, A. (2010). A genomic investigation of the ripening regulation in peach fruit. PhD thesis, University of Padua, Italy.

Tatsuki, M., Haji, T. and Yamaguchi, M. (2006). The involvement of 1-aminocyclopropane-1-carboxylic acid synthase isogene, Pp-ACS1, in peach fruit softening. *Journal of Experimental Botany* 57: 1281–9.

Trainotti, L., Zanin, D. and Casadoro, G. (2003). A cell wall-oriented genomic approach reveals a new and unexpected complexity of the softening in peaches. *Journal of Experimental Botany* 54: 1821–32.

Trainotti, L., Bonghi, C., Ziliotto, F., Zanin, D., Rasori, A. et al. (2006). The use of microarray μPEACH1.0 to investigate transcriptome changes during transition from pre-climacteric to climacteric phase in peach fruit. *Plant Science* 170: 606–13.

Trainotti, L., Tadiello, A. and Casadoro, G. (2007). The involvement of auxin in the ripening of climacteric fruits comes of age: the hormone plays a role of its own and has an intense interplay with ethylene in ripening peaches. *Journal of Experimental Botany* 58: 3299–308.

Trainotti, L., Cagnin, S., Forcato, C., Bonghi, C., Dhingra, A. et al. (2012). Functional genomics: transcriptomics. In: Kole, C. and Abbott, A.G. (Eds) *Genetics, Genomics and Breeding of Stone Fruits: Genetics, Genomics and Breeding of Crop Plants* (CRC Press, Taylor & Francis), pp. 292–322.

Vecchietti, A., Lazzari, B., Ortugno, C., Bianchi, F., Malinverni, R. et al. (2008). Comparative analysis of expressed sequence tags from tissues in ripening stages of peach (Prunus persica L. Batsch). *Tree Genetics & Genomes* 5: 377–91.

Verde, I., Vendramin, E., Micali, S., Dettori, M.T. and Sosinski, B. (2012). Genome sequencing initiative. In: Kole, C. and Abbott, A.G. (Eds) *Genetics, Genomics and Breeding of Stone Fruits: Genetics, Genomics and Breeding of Crop Plants* (CRC Press, Taylor & Francis), pp. 244–69.

Visai, C. and Vanoli, M. (1997). Volatile compound production during growth and ripening of peaches and nectarines. *Scientia Horticulturae* 70: 15–24.

Vizoso, P., Meisel, L.A., Tittarelli, A., Latorre, M., Saba, J. et al. (2009). Comparative EST transcript profiling of peach fruits under different post-harvest conditions reveals candidate genes associated with peach fruit quality. *BMC Genomics* 10: 423.

Vizzotto, G., Pinton, R., Varanini, Z. and Costa, G. (1996). Sucrose accumulation in developing peach fruit. *Physiologia Plantarum* 96: 225–30.

Wang, A., Yamakake, J., Kudo, H., Wakasa, Y., Hatsuyama, Y. et al. (2009). Null mutation of the MdACS3 gene, coding for a ripening-specific 1-Aminocyclopropane-1-Carboxylate synthase, leads to long shelf life in apple fruit. *Plant Physiology* 151: 391–9.

Watkins, C.B. (2006). The use of 1-methylcyclopropene (1-MCP) on fruits and vegetables. *Biotechnology Advances* 24: 389–409.

Whitford, R., Fernandez, A., Tejos, R., Pérez, A.C., Kleine-Vehn, J. et al. (2012). GOLVEN secretory peptides regulate auxin carrier turnover during plant gravitropic responses. *Developmental Cell* 22: 678–85.

Yamagami, T., Tsuchisaka, A., Yamada, K., Haddon, W.F., Harden, L.A. and Theologis, A. (2003). Biochemical diversity among the 1-Amino-cyclopropane-1-Carboxylate synthase isozymes encoded by the Arabidopsis gene family. *Journal of Biological Chemistry* 278: 49 102–12.

Ziliotto, F., Begheldo, M., Rasori, A., Bonghi, C. and Tonutti, P. (2008). Transcriptome profiling of ripening nectarine (*Prunus persica* L. Batsch) fruit treated with 1-MCP. *Journal of Experimental Botany* 59: 2781–91.

Ziosi, V., Bonghi, C., Bregoli, A.M., Trainotti, L., Biondi, S. et al. (2008). Jasmonate-induced transcriptional changes suggest a negative interference with the ripening syndrome in peach fruit. *Journal of Experimental Botany* 59: 563–73.

10

Application of doubled haploid technology in breeding of *Brassica napus*

Natalija Burbulis, Aleksandras Stulginskis University, Lithuania, and Laima S. Kott, University of Guelph, Canada

DOI: 10.1533/9781908818478.183

Abstract: *Brassica* is the most economically important genus in the *Brassicaceae* family. Among *Brassica* crops, oilseeds have the highest economic value, reflected in the fatty acid composition of the oil. Resistance to fungal pathogens and insect pests is frequently investigated in the selection of *Brassicas*. Cultivars resistant to *Sclerotinia* would be desirable for both ecological and economic reasons. Cabbage seedpod weevil and root maggot could be controlled by introgression of resistance genes and analysis of qualitative traits. Haploids of rapeseed are used to produce haploid calli for *in vitro* mutation selection. Homozygous rapeseed plants are generated by culturing microspores from flower buds of F_1 plants. The doubled haploid production system enables breeders to develop homozygous genotypes from heterozygous parents in one generation. Since the development of 'canola', rapeseed low in erucic acid and glucosinolates, research has shifted towards reduction of linolenic acid. Finally, selection methods for improved seed meal and cold tolerance are introduced, and the use of interspecific crosses is discussed.

Key words: brassica, pest resistance, qualitative trait loci (QTL), molecular marker linkage, double haploid (DH), oil composition, high performance liquid chromatography (HPLC), random amplified polymorphic DNA (RAPD) marker, salicylic acid, jasmonic acid, proline, cold tolerance.

10.1 Introduction

Brassica is the most economically important genus in the *Brassicaceae* family (syn. *Cruciferae*). Among *Brassica* crops, oilseeds have the highest economic value. When *Brassica* oils are low in aliphatic glucosinolates and erucic acid, the varieties are increasingly commonly referred to as canola. Canola, which is most often *Brassica napus*, has received much attention worldwide and may soon be the most popular oilseed crop.

The remarkable discovery that haploid embryos and plants can be produced by culturing anthers of *Datura* (Guha and Maheshwari, 1964) brought renewed interest in haploidy. Studies in haploid production were rapidly initiated in many species, and embryogenesis through anther and microspore culture has been reported in 247 species belonging to 88 genera and 34 families of angiosperms (Maheshwari et al., 1982). For *Brassica* species the first attempt at producing Double Haploid (DH) lines was made using the anther culture system (Thomas and Wenzel, 1975; Keller and Armstrong, 1977). However, subsequent research showed that in *Brassica* higher frequencies of regenerated plants can be achieved by isolated microspore culture (Chuong and Beversdorf, 1985; Swanson et al., 1987; Polsoni et al., 1988). Not only is microspore culture up to ten times more efficient in embryo yield than anther culture (Siebel and Pauls, 1989), but it also results in far fewer tetraploids and other genetic anomalies. Furthermore, microspore culture can easily be combined with other biotechnological methods, such as mutagenesis or gene transfer, in order to create novel genetic variation (Kott, 1998). Molecular geneticists, biotechnologists and plant breeders are currently further improving and modifying the quality of rapeseed oil and meal utilizing emerging haploid technologies. In this chapter, the application of microspore technology in rapeseed breeding is reviewed.

10.2 Technique of isolated microspore culture

Haploids of rapeseed are produced in the laboratory from gametic tissue, namely immature pollen grains or microspores. After Lichter (1982) had first successfully initiated a few haploid embryos in an enriched liquid medium from isolated microspores of *Brassica napus*, numerous researchers participated in the development of highly efficient haploid embryogenic systems (Pechan and Keller, 1988; Charne and Beversdorf, 1988; Mathias, 1988; Kott and Beversdorf, 1990). In general, the procedure can be subdivided into four major steps: 1) microspore extraction; 2) *in vitro* embryogenesis; 3) *in vitro* embryo germination; and 4) chromosome doubling. A protocol for doubled haploids production of *Brassica napus* (Fletcher et al., 1998) involves the following steps:

1. Unopened flower buds at the appropriate stage of development are harvested, surface-sterilized in 5.6% sodium hypochlorite bleach for 10–15 min. and rinsed in sterile distilled water three times. Microspores in the uninucleate stage are embryogenic (Kott et al., 1988).
2. Microspores are isolated by mechanical homogenization in 13% sucrose solution. The homogenate is passed through two nylon filters prior to centrifugation several times with fresh sterile medium.
3. Spores are suspended in liquid Nitsch Liquid Nutrient (NLN) medium at a density of 80–100000 spores ml^{-1} in Petri dishes. The dishes are sealed with a double layer of Parafilm and placed at 30°C in darkness for ten days followed by 18 days on a gyratory shaker (60 rpm) in the dark.
4. The mature cotyledonary embryos are transferred to plates of solid B_5 medium (Gamborg et al., 1968; modified with 1% sucrose), sealed and placed in a 4°C chamber for ten days.
5. After cold exposure, plates are moved to 25°C in light for germination for approximately 30 days and then haploid plantlets are transplanted to soil and placed in the greenhouse.
6. When plants begin to flower, the chromosome number of haploids is doubled with 0.34% colchicine solution.

10.3 Doubled haploid method in breeding of *Brassica napus*

The doubled haploid production system is a protocol used to generate homozygous rapeseed plants by culturing microspores isolated from young flower buds of F_1 plants. The DH system is an extremely valuable breeding tool because it enables breeders to develop completely homozygous genotypes from heterozygous parents in one single generation. Every microspore-derived embryo represents a unique combination of traits from each parent in the original cross. Doubled haploids allow the genes of recombinant gametes to be fixed directly as fertile homozygous lines. Complete homozygosity is achieved very rapidly, since DH seed is harvested only 8–9 months after culture for spring types and 1.5 years for winter types (Kott, 1998). The value of doubled haploids in breeding is that they are diploid, fertile and homozygous for all traits, and therefore true breeding in the following generations. Furthermore, because the DH system utilizes uninucleate microspores as the starting material, the protocol can be variously manipulated to serve many research and breeding purposes. Major advantages of this method in comparison to conventional breeding methods via repeated self-pollination include reduction of time and space for breeding and ultimately reduction of the costs for cultivar development.

The doubled haploid production system via microspore culture is of immeasurable value for crop breeding in conjunction with other *in vitro* technologies such as mutagenesis and selection.

10.4 *In vitro* mutagenesis

Although random mutations occur naturally in spore populations, the incidence of genetically altered genotypes can be significantly increased by the application of mutagens to haploid cells. The optimal time for *in vitro* application of a mutagenizing agent in isolated microspore cultures is at the single cell stage, shortly after extraction. This ensures homogeneity of embryo tissues, since all cell nuclei within the developing embryo subsequently will have been derived from the original mutagenized cell and chimeras are avoided. This culture system is ideal for crop improvement using *in vitro* mutagenesis because:

- Haploids and doubled haploids express all mutations, which can be recognized during selection.
- Availability of large microspore/embryo populations increases the probability of identifying beneficial mutants. Large cultures used for *in vitro* mutagenesis consist of approximately 200 culture plates, each with 1 million microspores, which in total can yield well over 2 million mutagenized embryos (Polsoni et al., 1988). When large cultures are produced several times a week, the probability of identifying a desired genetic variance increases.
- *In vitro* mutagenesis is fast and clean.

The two types of mutagens that have been used successfully in microspore culture to generate genetic variants in *Brassica* are chemical mutagens such as sodium azide (NaN_3), ethylnitrosourea (ENU) and ethylmethanesulfonate (EMS) and radioactive mutagens, gamma and ultraviolet radiation (Beversdorf and Kott, 1987; Polsoni et al., 1988; Swanson et al., 1988; Swanson et al., 1989). If recommended precautions are followed, radioactive mutagens are faster, cleaner and safer to use in comparison to chemical mutagens, primarily because spores are extremely fragile during the first few days of culture when mutagenic chemicals are applied and the subsequent screening or centrifuging of cultures to remove residual chemicals can easily damage swollen spores or dividing cells. Second, radioactive mutagens are better simply because handling of dangerous carcinogenic compounds is unnecessary. To ensure a consistent and predictable level of dose intensity for genetic alterations in microspore culture, a kill curve is generated. The current mutation breeding technology using milder mutagens, such as UV radiation, induces nucleotide base substitutions (point mutations) without causing major physical damage to chromosome structure. In 2005, over 2300 officially released mutant varieties were cited in the Food and Agriculture Organization (FAO)/International Atomic Energy Agency (IAEA) Mutant Varieties Database (Jain, 2005).

The protocol for *in vitro* mutagenesis and selection is the same as for microspore culture with two additional steps, namely *in vitro* mutagenesis at Day 1 and *in vitro* selection (Kott, 1998). The microspore culture system is most effective in plant breeding if both the mutagen application and the selection of novel traits can be carried out in vitro. The classic example of *in vitro* mutagenesis and selection is for herbicide resistance, where the active chemical is introduced into the culture medium after mutation treatment. Several mutant rapeseed lines with resistance to herbicides have been developed using this system (Beversdorf and Kott, 1987; Swanson et al., 1988; Polsoni et al., 1988; Swanson et al., 1989). Selection for herbicide resistance is likely to be the simplest test for this system, since a single point mutation at any one of a number of genes could effectively interfere with the uptake, assimilation or translocation of the herbicide, resulting in a resistant/tolerant plant. Nevertheless, mutation technology coupled with embryogenesis in microspore culture of rapeseed is currently being exploited for selection of other traits, for example, an improvement in the seed quality by modifying the fatty acid profiles and altering the glucosinolate and sinapine level in the oil and meal. The multigenic control of these qualitative traits makes *in vitro* selection more costly and difficult.

10.5 Utilization of double haploidy in selection for resistance

10.5.1 Resistance to cabbage seedpod weevil

The cabbage seedpod weevil, *Ceutorhynchus obstrictus* (Marsham), is an insect pest of major economic importance in the production of canola (*Brassica napus* L. and *Brassica rapa* L.) in Europe and North America. *Ceutorhynchus obstrictus* can cause economic losses to rapeseed at several stages of crop development; therefore, the benefits to canola breeders and the industry of developing weevil-resistant germplasm would include reduced input costs, pesticide use and environmental degradation, and increased yield. Yellow mustard (*Sinapis alba* L.) is resistant to *C. obstrictus* (cabbage seedpod weevil; CSPW), although the exact mechanism is not known (McCaffrey et al., 1999). A unique canola population was generated at the University of Guelph from a cross between *B. napus* and *S. alba* (Kott and Dosdall, 2004). Intergeneric lines were developed by performing crosses between *S. alba* cv. Kirby and *B. napus* cv. Topas to produce F_1 hybrids using embryo rescue technology (Ripley and Arnison, 1990). The F_1 hybrids were then backcrossed with the *B. napus* parent for three generations. A total of 563 doubled haploids were produced from 12 of these lines, by the *in vitro* method of Fletcher et al. (1998), and 230 lines were subsequently selected from the doubled haploids for having canola-quality seed glucosinolate levels (Dosdall and Kott, 2006) and were used for further breeding. The hypothesis was that some DH progeny from this cross inherited resistance to CSPW from *S. alba*. Weevil infestation levels were assessed for the *B. napus* × *S. alba* BC2 and BC3 DH populations in the field over 7 years in Alberta, where weevil pressure is strong, to establish the resistant or susceptible status of these lines (Shaw et al., 2009). The basic objectives for this study were to confirm field resistance in the *B. napus* × *S. alba* germplasm in Ontario and to identify any biochemical markers associated with resistance/susceptibility. Canola doubled haploid lines derived from BC2 or BC3 families were field screened for resistance followed by chemical analysis of glucosinolates to detect biochemical polymorphisms correlated with CSPW resistance using high performance liquid chromatography (HPLC). Two polymorphic peaks were found, one each from extracts of upper cauline leaves and Stage 3 pod seed, with retention times of ~23 and 19 min., respectively. These HPLC peaks consistently correlated with larval infestation data, and the peak differences between R and S double haploid (DH) lines were significant. Therefore, the authors suggested that these two peaks can be used as biochemical markers in breeding programs and may play a role in rapid and early detection of CSPW resistance.

10.5.2 Resistance to root maggot

One of the major goals of sustainable agriculture is to reduce the use of agricultural pesticides and to develop integrated pest management (IPM) programs. Since some root maggot resistance has been found in members of the *Brassicaceae* (Ellis et al., 1999; Jyoti et al., 2001; Jensen et al., 2002), an important component of an IPM program in this agroecosystem is to breed rapeseed genotypes with root maggot resistance. Crosses between white mustard (*Sinapis alba* cv. Kirby) and rapeseed (*Brassica napus* cv. Topas) were initiated to develop F_1s using embryo rescue technology; the F_1s were then crossed again with *B. napus* cvs Topas and Westar to the backcross 3 (BC3) generation (Kott and Dosdall, 2004). From 12 BC3 lines, a total of 563 doubled haploids were produced by the standard *in vitro* method. Approximately half of these BC3 DH lines were of canola quality for glucosinolate content. Canola-quality DH lines were field-screened for root maggot resistance, and those with best resistance were used for crossing with elite breeding material in the University of Guelph canola breeding program, to introgress the resistance trait into genotypes with improved agronomic quality. The 212 new DHs produced from these crosses were field-screened under heavy root maggot pressure and the DHs with best root maggot resistance scores were tested for oil quality. Candidate DHs with the best oil quality traits were again used to introgress the root maggot resistance trait into new canola genotypes in the breeding program. Field evaluation determined that the vast majority of the DH population had mean root maggot damage values that were substantially lower than *B. napus* cv. Q2, the check genotype. Subsequently developed DH lines were used to identify quantitative trait loci for resistance to root maggot damage (Ekuere et al., 2005). Two quantitative trait loci (QTL) on a molecular marker linkage map that explain 55% of the variance for this trait were identified. The example clearly shows how the double haploid system can rapidly fix desired traits, simplifying subsequent selection in the field.

10.5.3 Resistance to Sclerotinia sclerotiorum

Sclerotinia sclerotiorum (Lib.) de Bary is a necrotrophic fungal pathogen that infects many plant species, including *Brassica napus*. Several chemicals have been used to control this disease; however, the fungicides are often ineffective and not environmentally safe. Therefore, cultivation of rapeseed cultivars highly resistant to *Sclerotinia* would be most desirable for both ecological and economic reasons.

Theoretically, if *Brassica napus* plants were able to disarm fungal infections by enzymatic breakdown of the fungal acids that destroy leaf mesophyll, fungus damage could be largely blocked. *Sclerotinia* infection uses oxalic acid to advance lesions in leaves and stems, and therefore, using oxalic acid as a selection agent in cultures of UV mutagenized microspores, it has been possible to identify embryos *in vitro* that are able to overcome high levels of acid (Kott et al., 2002). Most likely the defence mechanism is the production of an oxidase that breaks down this acid. The method used was as follows: short UV radiation generated point mutations in microspores immediately after isolation at the LD_{50} level. After 14 days in culture, the selection agent, oxalic acid, was added to the medium at a final concentration of 0.5 mM. Most embryos died, and survivors were carried through to chromosome doubling and seed production. However, at the haploid stage, plantlets were again screened using the 'leaf wilt test', in which one or two leaves were severed from each plant and the cut ends were placed in oxalic acid at levels ranging from 80 to 200 mM. Leaves that remained turgid after 18 hours rapidly identified lines that were strongly resistant. Further testing with the pathogen in a controlled environment eliminated any further escapes. DH canola lines produced in this manner have performed very well in official field trials. For example, in the North Dakota Sclerotinia Canola Variety Trial in 2005, of 28 mostly commercial entries, four of the *in vitro* selected lines scored first, third, fourth and ninth for percentage Incidence of Disease (DI), with scores of 4.5, 9.5, 11.0 and 15.0, respectively. The mean DI for this trial was 27.0 and the range from 4.5 to 68.0%. Clearly, using other appropriate acids, this *in vitro* selection method could be used to develop tolerance or resistance to other fungal pathogens.

Liu et al. (2005) used haploid seedlings derived from microspore cultures to produce haploid calli for *in vitro* mutation selection. Mutation was induced by ethylmethane sulfonate or occurred spontaneously, and oxalic acid, as a selection agent, was added to the medium for screening of resistant mutants. Of the 54 DH lines produced from the *in vitro* mutation selection, two DH lines of resistant mutants, named M083 and M004, were selected following seedling and glasshouse tests.

10.6 Selection for modified seed oil composition

The commercial value of the oilseed *Brassicas* is reflected in the fatty acid composition of the edible oil. Rapeseed oil has the lowest level of saturated fatty acids (7%) of any vegetable oil, a desirable high level of both mono-unsaturated fatty acid (60%) and omega-3 fatty acid (10%), and a reasonable amount of essential fatty acids (20%). Recent nutritional studies indicate that oleic acid reduced the level of harmful low density lipoprotein cholesterol in humans. From the human health perspective, rapeseed oil could bring further benefit by elevating the mono-unsaturated oleic fatty acid content and by reducing the levels of the saturated fatty acids (stearic, palmitic) and the poly-unsaturated linolenic fatty acid. Increased industry demand and utilization of speciality oils have renewed the interest of researchers and plant breeders, and the microspore culture system has already been utilized in accelerated development of new oil products through mutagenesis (Wong and Swanson, 1991). The suitability of vegetable oil for human consumption or use in industrial processing is determined by the biochemical properties and concentrations of the fatty acids comprising the oil. Palmitic (16:0), stearic (18:0), oleic (18:1), linoleic (18:2) and linolenic (18:3) acids dominate the fatty acid profile of rapeseed seed, each being at a concentration greater than 1% of the total fatty acids, while the remaining ten fatty acids (12:0, 14:0, 16:1, 20:0, 20:1, 20:2, 22:0, 22:1, 24:0, 24:1) are typically detected at levels less than 1%. The saturated fatty acids in rapeseed oil include lauric, myristic, palmitic, stearic, arachidic, behenic and lignoceric acids. Rapeseed/canola oil is promoted as the healthiest vegetable oil for human consumption due to its low levels of saturated fatty acids (ca. 7%) and high levels of mono-unsaturated oleic acid in addition to moderate levels of omega-3 and six fatty acids (Ackman, 1990).

Since the development of 'canola', which is rapeseed low in both erucic acid and glucosinolates, oil quality research has shifted towards the reduction of linolenic acid (18:3). Linolenic acid is a trienoic acid, which is easily oxidized to cause an off flavour and odour of the oil, resulting in a shortened shelf life. Rapeseed cultivars usually contain 8–12% linolenic acid in the oil, and partial hydrogenation of rapeseed oil is typically used to lower the amount of 18:3. This process, however, results in the production of *trans* fatty acids that are detrimental to human health. Therefore, the development through breeding of rapeseed cultivars with lower 18:3 is considered more desirable. Low linolenic acid rapeseed genotypes have been developed through chemical mutagenesis (Röbbelen and Thies, 1980) and interspecific hybridization (Roy and Tarr, 1985); however, the breeding efforts are complicated by maternal effects, multiple gene inheritance and large environmental effects on the fatty acid contents in rapeseed oil (Chen and Beversdorf, 1990; Diepenbrock and Wilson, 1987; Kondra and Thomas, 1975; Thomas and Kondra, 1973). Several research groups have reported the identification of molecular markers associated with linolenic acid level in *B. napus*. Tanhuanpää et al. (1995) identified one random amplified polymorphic DNA marker (RAPD) associated with a gene controlling 18:3 level, which accounted for 23% of the genetic variability in the trait, and similarly Hu et al. (1995) found one restriction fragment length polymorphism marker (RFLP) that accounted for 26% of the phenotypic variation in the trait. Thormann et al. (1996) identified two QTL on a molecular marker linkage map that accounted for 60% of the variance in linolenic acid content. Two QTL which explained 24% and 30.7% of the variance in the trait were also mapped by Jourdren et al. (1996a). Two independent RAPD markers accounted for a total of 39% of the genetic variability for 18:3 in a doubled haploid population (Rajcan et al., 1999). It was found that two major loci controlling linolenic acid content correspond to the two *fatty acid desaturase 3* (*fad3*) genes (Jourdren et al., 1996b; Barret et al., 1999). Somers et al. (1998) identified three QTL that collectively explained 51% of the phenotypic variation for linolenic acid content within the DH population tested and found that the *fad3* gene localized near one of the QTL. Schierholt et al. (2001) identified an association between the *fad2* gene and oleic acid desaturation. Hu et al. (2006) developed two single nucleotide polymorphism (SNP) markers, corresponding to the *fad2* and *fad3c* gene mutations. Authors suggest that these markers can be highly useful for direct selection of desirable *fad2* and *fad3c* alleles and breeding of rapeseed with high oleic and low linolenic acid.

The reduction of saturated fats in rapeseed/canola oil for commercial and human health benefits has recently been promoted as a goal for breeders. Many studies of seed oil formation have been performed with microspore-derived embryos, as it was found that such embryos and zygotic embryos accumulate triglycerides in a similar way (Pomeroy et al., 1991; Wiberg et al., 1991; Taylor and Weber, 1994). Microspore-derived embryos were analysed for fatty acid composition at different stages of development to determine the stage that closely resembled the fatty acid profile of an embryo from a mature seed. It was found that fatty acid composition was similar in the seed and in 28-day-old cultured embryos (Wong and Swanson, 1991). The protocol allows analysis of pieces of cotyledonary tissue from mutagenized haploid embryos to identify novel genotypes with improved oil quality without destruction of the embryo. Since this procedure preserved the embryo root/shoot axis, selected embryos were germinated and doubled, thus fixing the selected genotype. Canadian canola breeders used mutagenesis and an *in vitro* heat selection system for development of DH lines with major saturates (palmitic and stearic) reduced by several per cent (Beaith et al., 2005). Mutant embryos generated from direct ultraviolet radiation mutagenesis of microspores *in vitro* were subjected to heat during the maturation stage. Heat artificially elevated the saturate levels in developing mutant embryos, allowing efficient identification of those with reduced saturates within the expanded range using HPLC fatty acid analysis of the embryo cotyledons. Mutagenesis produced embryos with changes in fatty acids in both directions. Major saturate levels in the cotyledons of heat-treated mutant embryos ranged from 3.3 to 16.4% (heated control ca. 6–9%) and 1.3 to 10% (heated control ca. 2–4%) for palmitic and stearic fatty acids, respectively. Doubled haploid seed derived from embryos grown at normal temperatures confirmed the reduction of major saturates. HPLC fatty acid analysis of DH seed identified saturate levels ranging from 3.9 to 6.5% (control ca. 5.5%) and 0.9 to 2.7% (control ca. 1.7%) for palmitic and stearic fatty acids, respectively. Various doubled haploids were identified with major saturate levels below 5.5%.

10.7 Selection for improved seed meal

Members of the genus *Brassica* characteristically accumulate glucosinolates in seed and vegetative tissues. Glucosinolates are sulphur-based secondary metabolites that are damaging to proteins when hydrolysed by specific enzymes and are responsible for the pungent aroma associated with cole crops and mustards. The presence of these compounds in seed meal after crushing results in goitrogenic effects and reduced weight gain when fed to monogastric livestock, seriously reducing the value of rapeseed meal despite its high protein content. Low glucosinolate content constitutes one zero of the 'double zero' designation ('00'), along with reduced content of erucic acid, which together define cultivars of oilseed

Brassicas as 'canola'. Specifically, canola is a high-quality rapeseed with less than 1% erucic acid (22:1) in the seed oil and less than 15–30 µmol of glucosinolates per gram of meal. Germplasm with a tenfold decrease in alkenyl glucosinolate content has been developed through classical breeding since the discovery of a low-glucosinolate Polish cultivar, Bronowski (Stefansson and Kondra, 1975). However, Sharpe and Lydiate (2003) reported the difficulty of eliminating residual segments of the Bronowski genotype from elite material. A DH population of *B. napus* was analysed to identify molecular markers associated with glucosinolate content (Uzunova et al., 1995). In this study four quantitative trait loci that explain approximately 62% of the phenotypic variance of this trait were mapped. Further, a number of research groups have reported detection of QTL for total seed glucosinolate content in different rapeseed crosses (Howell et al., 2003; Sharpe and Lydiate, 2003; Zhao and Meng, 2003; Basunanda et al., 2007). Four quantitative trait loci on *B. napus* chromosomes were identified in a few studies; therefore it was assumed that these QTL represent major loci associated with seed glucosinolate content in different genotypes. Hasan et al. (2008) found evidence that known QTL for total seed glucosinolate content in *B. napus* might be associated with four genes involved in the biosynthesis of indole and aliphatic and aromatic glucosinolates. In a comparative study of glucosinolate metabolism in zygotic and microspore-derived embryos, it was found that glucosinolates appearing in cultured embryos are typical of those in zygotic embryos (Kott et al., 1990; McClellan et al., 1993). The *in vitro* mutagenesis and selection system has been utilized to rapidly and cheaply identify mutagenized haploid embryos with reduced total glucosinolate level (Kott, 1995; Burbulis et al., 2001).

Brassica napus seed is black or brown due to the accumulation of condensed tannins in the integument. Yellow-seeded rapeseed lines have been a long-term objective for many canola breeding programs, since it has been shown that yellow seed contains significantly more oil and protein and displays lower fibre content than black-seeded lines (Rashid et al., 1995; Simbaya et al., 1995). Numerous yellow-seeded *Brassica napus* lines have been developed through interspecific crosses with related yellow-seeded *Brassica* species (Hou-Li et al., 1991; Rahman, 2001; Rashid et al., 1994; Meng et al., 1998; Relf-Eckstein et al., 2003); however, only one yellow-seeded cultivar has been released to date (Liu et al., 1991). The introgression of genes encoding seed pigmentation from related *Brassica* species and subsequent expression of yellow seed colour in *Brassica napus* can be complex due to polyploidy, multiple gene control and maternal determination (Van Deynze et al., 1993; Tang et al., 1997). A new source of yellow-seeded *Brassica napus* canola has been identified in doubled haploid progeny, from a cross between the two black-seeded spring cultivars Star and Bolero (Burbulis and Kott, 2005). Six yellow-seeded doubled haploid lines were extracted from the F_1 generation of this cross. The yellow-seeded doubled haploid lines and some of their progeny showed differences in yellow seed colour expression in different temperature environments, among the 11 lines investigated. Results clearly show that high temperatures inhibit the accumulation of dark pigments in the seed coats of these genotypes. Line NL-350-1, which had yellow–brown and dark yellow seed development at 20 and 28°C, was the exception, producing brown seed at 30°C. This phenomenon may be related to colour expression reversion in this mutant which is brought on by extreme heat. Oil content of new *B. napus*-derived genotypes described herein is consistent with oil-enhanced yellow-seeded lines sourced from interspecific crosses. Tested doubled haploid lines showed a good range of oil content, ranging from 31.2 to 51.6% in the cooler environment, while best performance in the hot temperature reached 48–49% among the three best lines.

Doubled haploid populations were used to study the inheritance of seed colour in *Brassica napus*. A true-breeding *Brassica napus* line, DH268-2, and black-seeded cultivars, Star and Bolero, were used for reciprocal crosses from which DH populations were produced (Burbulis et al., 2006). In total 268 doubled haploid lines were produced that gave seed of sufficient quality to classify their colour. Self-pollinated seeds from DH lines from four crosses were grouped into seed colour classes as follows: black, brown, yellow. Chi-square goodness-of-fit test was used to compare the observed distribution in the segregating populations with those predicted by different models for seed colour inheritance (Shirzadegan, 1986; Van Deynze and Pauls, 1994; Baetzel et al., 2003). Depending on the source of yellow-seeded plants used in the genetic studies, in most cases a trigenic inheritance has been proposed. The segregation data obtained for the DH populations in our study was not consistent with the corresponding ratio for DHs predicted by these models. The discrepancy between the Burbulis (2005) and other studies may be due to differences in the genetic material used in the crosses. The yellow seed colour character in DH268-2 is derived from black-seeded cultivars Star and Bolero. It is assumed that the yellow seed trait in this line results from spontaneous mutations, in which mutant gene(s) block the synthesis of seed coat pigment (Burbulis, 2005). In order to create new yellow-seeded genotypes, an isolated microspore culture was used.

The donor plants for microspore culture were F_1 hybrids obtained from reciprocal crosses between the yellow-seeded rapeseed lines DH268-2 and DH268-20 and black-seeded cultivars Star, Bolero and Dynamite. The DH lines were selected due to the expression of the character 'yellow seeded' in the following generation. Following a phase of seed multiplication in the field, a large variation in seed characters of tested DH lines has been generated. Additionally, the selection of the DH lines and hybrids' progeny has continued, regarding different agronomic properties (Burbulis et al., 2007).

The sinapine content of rapeseed meal contributes to dark-coloured meal and to the off-flavour of eggs when it is fed to some poultry varieties. Although a number of different phenolics are involved in the oxidation reaction, the abundance of sinapine in *Brassica* meal suggests that this sinapoyl derivative is largely the cause of the phenolic browning process. Amar et al. (2008) studied 148 winter rapeseed DH lines with a large variation in sinapate ester content and mapped four QTL explaining 53% of genetic variance for this trait. The authors suggest that there is a pleiotropic effect of the two erucic acid genes on sinapate ester content; the effect of the alleles for low erucic acid content is to increase sinapate ester content. For further details on the improvement of meal quality in rapeseed using microspore culture, readers are referred to the reviews of Kott et al. (1990, 1996).

10.8 Selection for cold tolerance

Hawkins et al. (2002) studied the relationship between vernalization requirement and freezing tolerance using a spring-type rapeseed doubled haploid line (Vern-) derived from a cross between winter rapeseed cultivars with different freezing tolerance and vernalization requirements, and found that this line lacked a vernalization requirement and possessed a high freezing tolerance. Moreover, it was found that the two traits could be separated and positive trait expressions could be combined. The authors suggested that the (Vern-) line is a very valuable resource for a breeding program to bridge spring and winter gene pools.

Furthermore, another study has been conducted using microspore culture in order to develop spring canola lines with mutations with altered biochemical pathways that increase cold tolerance (McClinchey and Kott, 2008). The approach was to generate UV point mutations in cultured microspores followed by chemical *in vitro* selection of individual mutant microspores or embryos resulting in measurable alterations to various biochemical pathways with elevated levels of key defence signalling molecules such as salicylic acid (SA), p-fluoro-D,L-phenylalanine (FPA) and jasmonic acid (JA). In addition, since proline (Pro) is known to protect plant tissues in the cold-induced osmotic stress pathway, mutants that overproduce Pro were selected *in vitro* by using three Pro analogues: hydroxyproline (HP), azetidine-2-carboxylate (A2C), and 3,4-dehydro-D,L-proline (DP). Of the 329 *in vitro* selected mutant embryos produced, 74 were identified with significant cold tolerance compared with their donor parents through indoor freezer tests at −6°C, and 19 had better winter field survival than winter canola checks. All chemically selected mutant doubled haploids with increased cold tolerance compared well with parent lines for all seed quality and agronomic parameters.

10.9 Concluding remarks

Lessons learned over the past four decades in the manipulation of cells, tissues and organs in the *Brassica* genus have led us to an impressive point in cultivar and trait development for the improvement of canola for human consumption and commercialization. Doubled haploidy, together with *in vitro* mutagenesis and selection, has simplified the alteration of the chemical makeup of canola oil and has allowed researchers to easily modify oil profiles (reduction of erucic acid and saturates, increase of oleic acid), as well as demystifying the reduction of glucosinolates and sinapine in oil and seed meal to benefit both human and livestock health. Furthermore, doubled haploidy has played a major part in rapid development of genotypes resistant to pests (diseases, insects) by fixing introgressed genes and allowing direct *in vitro* selection for resistance. Obviously, these genotypes require less application of agrochemicals, producing healthier canola products for human consumption that are less costly to produce. Clearly, the experimental studies presented in this chapter show that the microspore culture system in conjunction with other *in vitro* techniques is an extremely powerful tool for the development of improved *Brassica napus* cultivars.

10.10 References

Ackman R.G. 1990. Canola fatty acids – an ideal mixture for health, nutritional and food use. In: Shahidi F. (ed.) *Canola and Rapeseed: Production, Chemistry,*

Nutrition and Processing Technology. Van Nostrand Reinhold, New York. pp. 81–90.

Amar S., Ecke W., Becker H.C., Möllers C. 2008. QTL for phytosterol and sinapate ester content in *Brassica napus* L. collocate with the two erucic acid genes. *Theor. Appl. Genet.* 116: 1051–61.

Baetzel R., Luhs W., Badani A.G., Friedt W. 2003. Development of segregating populations in the breeding of yellow-seeded winter rapeseed (*Brassica napus* L.). In: *Proceeding of the 11th International Rapeseed Congress.* Copenhagen, Denmark. pp. 238–42.

Barret P., Delourme R., Brunet D., Jourdren C., Horvais R., et al. 1999. Low linolenic acid level in rapeseed can be easily assessed through the detection of two single base substitution in fad3 genes. In: *Proceeding of the 10th International Rapeseed Congress.* Canberra, Australia. pp. 26–9.

Basunanda P., Spiller T.H., Hasan M., Gehringer A., Schondelmaier J., et al. 2007. Marker-assisted increase of genetic diversity in a double-low seed quality winter oilseed rape genetic background. *Plant Breeding* 126: 581–7.

Beaith M.E., Fletcher R., Kott L.S. 2005. Reduction of saturated fats by mutagenesis and heat selection in *Brassica napus* L. *Euphytica* 144: 1–9.

Beversdorf W.D., Kott L.S. 1987. An *in vitro* mutagenesis/selection system for *Brassica napus*. *Iowa State J. Res.* 61: 435–43.

Burbulis N. 2005. The inheritance of seed colour in spring rapeseed doubled haploid populations. *Vagos* 67: 7–11.

Burbulis N., Malinauskaite R., Kott L. 2001. Oil quality improvement through *in vitro* mutagenesis and haploid selection in Lithuanian winter *Brassica napus* germplasm. *Proc. Latv. Acad. Sci.* 55: 197–200.

Burbulis N., Kott L.S. 2005. A new yellow-seeded canola genotype originating from double low black-seeded *Brassica napus* cultivars. *Can. J. Plant Sci.* 85: 109–14.

Burbulis N., Kuprienė R., Žilėnaitė L., Ramaškevičienė A. 2006. The evaluation and modification of genetic factors, regulating synthesis of pigments in spring rapeseed. *Sodininkystė ir daržininkystė* 25: 74–83.

Burbulis N., Kuprienė R., Blinstrubienė A., Juozaitytė R., Žilėnaitė L. 2007. Application of biotechnology methods for spring rapeseed (*Brassica napus* L.) breeding. *Žemdirbystė=Agriculture* 94: 129–38.

Charne D.G., Beversdorf W.D. 1988. Improving microspore culture as a rapeseed breeding tool: the use of auxins and cytokinins in an induction medium. *Can. J. Bot.* 66: 1671–5.

Chen J., Beversdorf W.D. 1990. Fatty acid inheritance in microspore-derived populations of spring rapeseed (*Brassica napus* L.). *Theor. Appl. Genet.* 80: 465–9.

Chuong P.V., Beversdorf W.D. 1985. High frequency embryogenesis through isolated microspore culture in *Brassica napus* L. and *B. carinata* Braun. *Plant Sci.* 39: 219–26.

Diepenbrock W., Wilson R.F. 1987. Genetic regulation of linolenic acid concentration in rapeseed. *Crop Sci.* 27: 75–7.

Dosdall L.M., Kott L.S. 2006. Introgression of resistance to cabbage seed pod weevil to canola from yellow mustard. *Crop Sci.* 46: 2437–45.

Ekuere U.U., Dosdall L.M., Hills M., Keddie A.B., Kott L., et al. 2005. Identification, mapping and economic evaluation of QTLs encoding root maggot resistance in *Brassica*. *Crop Sci.* 45: 371–8.

Ellis P.R., Pink D.A.C., Mead A. 1999. Identification of high levels of resistance to cabbage root fly, *Delia radicum*, in wild *Brassica* species. *Euphytica* 110: 207–14.

Fletcher R., Coventry J., Kott L.S. 1998. *Doubled Haploid Technology for Spring/Winter Brassica Napus*. OAC Publication. University of Guelph, Guelph, Ontario, Canada.

Gamborg O.L., Miller R.A., Ojiwa K. 1968. Nutrient requirements of suspension culture of soybean root callus. *Exp. Cell Res.* 50: 151–8.

Guha S., Maheshwari S.C. 1964. In vitro production of embryos from the anthers of *Datura*. *Nature* 204: 497.

Hasan M., Friedt W., Pons-Kühnemann J., Freitag N.M., Link K., et al. 2008. Association of gene-linked SSR markers to seed glucosinolates content in oilseed rape (*Brassica napus* ssp. *napus*). *Theor. Appl. Genet.* 116: 1035–49.

Hawkins G.P., Deng Z., Kubik T.J., Johnson-Flanagan A.M. 2002. Characterization of freezing tolerance and vernalization in Vern-, a spring-type *Brassica napus* line delivered from a winter cross. *Planta* 216: 220–6.

Hou-Li L., Han J.X., Hu X.J. 1991. Studies on the inheritance of seed coat colour and other related characteristics of yellow seeded *Brassica napus*. In: *Proceeding of the 8th International Rapeseed Congress*. Saskatoon, Canada. pp. 1438–44.

Howell P.M., Sharpe A.G., Lydiate D.J. 2003. Homoeologous loci control the accumulation of seed glucosinolates in oilseed rape (*Brassica napus*). *Genome* 46: 454–60.

Hu J., Quiros C., Arus P., Struss D., Röbbelen G. 1995. Mapping of a gene determining linolenic acid concentration in rapeseed with DNA-based markers. *Theor. Appl. Genet.* 90: 258–62.

Hu X., Sullivan-Gilbert M., Gupta M., Thompson S.A. 2006. Mapping of the loci controlling oleic and linolenic acid contents and development of *fad2* and *fad3* allele-specific markers in canola (*Brassica napus* L.). *Theor. Appl. Genet.* 113: 497–507.

Jain M.S. 2005. Major mutation-assisted plant breeding programs supported by FAO/IAEA. *Plant Cell Tiss. Org.* 82: 113–23.

Jensen E.B., Felkl G., Kristiansen K., Andersen S.B. 2002. Resistance to the cabbage root fly, *Delia radicum*, within *Brassica fruticulosa*. *Euphytica* 124: 379–86.

Jourdren C., Barret P., Horvais R., Delourme R., Renard M. 1996a. Identification of RAPD markers linked to linolenic acid genes in rapeseed. *Euphytica* 90: 351–7.

Jourdren C., Barret P., Brunel D., Delourme R., Renard M. 1996b. Specific molecular marker of the genes controlling linolenic acid content in rapeseed. *Theor. Appl. Genet.* 93: 512–18.

Jyoti J.L., Shelton A.M., Earle E.D. 2001. Identifying sources and mechanisms of resistance in crucifers for control of cabbage maggot (*Diptera: Anthomyiidae*). *J. Econ. Entomol.* 94: 942–9.

Keller W.A., Armstrong K.C. 1977. Embryogenesis and plant regeneration in *Brassica napus* anther culture. *Can. J. Bot.* 55: 1383–8.

Kondra Z.P., Thomas P.M. 1975. Inheritance of oleic, linoleic and linolenic acids in seed oil of rapeseed (*Brassica napus*). *Can. J. Plant Sci.* 55: 205–10.

Kott L. 1995. Production of mutants using the rapeseed doubled haploid system. In: *Proceeding of an International Symposium on the Use of Induced Mutations and Molecular Techniques for Crop Improvement*. International Atomic Energy Agency, Vienna. pp. 505–15.

Kott L.S. 1998. Application of doubled haploid technology in breeding of oilseed *Brassica napus*. *AgBiotech News and Information* 10: 69–74.

Kott L.S., Beversdorf W.D. 1990. Enhanced plant regeneration from microspore-derived embryos of *Brassica napus* by chilling, partial desiccation and age selection. *Plant Cell Tiss. Org.* 23: 187–92.

Kott L.S., Dosdall L.M. 2004. Introgression of root maggot resistance (*Delia* spp.) derived from *Sinapis alba* L. into *Brassica napus* L. *Brassica* 6: 55–62.

Kott L.S., Polsoni L., Beversdorf W.D. 1988. Cytological aspects of isolated microspore culture of *Brassica napus*. *Can. J. Bot.* 66: 1658–64.

Kott L.S., Erickson L.R., Beversdorf W.D. 1990. The role of biotechnology in canola/rapeseed research. In: Shahidi F. (ed.) *Canola and Rapeseed: Production, Chemistry, Nutrition and Processing Technology*. Van Nostrand Reinhold, New York. pp. 47–78.

Kott L.S., Wong R., Swanson E., Chen J. 1996. Mutation and selection for improved oil and meal quality in *Brassica napus* utilizing microspore culture. In: Jain S.M., Sopory S.K., Veilleux R.E. (eds) *In Vitro Haploid Production in Higher Plants*. Kluwer, Dordrecht. pp. 151–67.

Kott L., Kyrychenko I., Fletcher R. 2002. *Sclerotinia* resistance in *Brassica napus* derived from *in vitro* UV-induced mutations. In: *International Conference Biotechnology Approaches for Exploitation and Preservation of Plant Resources*. Yalta, Ukraine. pp. 84–5.

Lichter R. 1982. Induction of haploid plants from isolated pollen of *Brassica napus*. *Z. Pflanzenzuchtung* 105: 427–34.

Liu H.L., Han J.X., Hu X.J. 1991. Studies of the inheritance of seed coat colour and other related characteristics of yellow seeded *Brassica napus*. In: *Proceeding of the 8th International Rapeseed Congress*. Saskatoon, Canada. pp. 1438–44.

Liu S., Wang H., Zhang J., Fitt B.D.L., Xu Z., et al. 2005. *In vitro* mutation and selection of doubled-haploid *Brassica napus* lines with improved resistance to *Sclerotinia sclerotiorum*. *Plant Cell Rep.* 24: 133–44.

Maheshwari S.C., Rashid A., Rayagi A.K. 1982. Haploids from pollen grains – retrospect and prospect. *Am. J. Bot.* 69: 865–79.

Mathias R. 1988. An improved *in vitro* culture procedure for embryoids derived from isolated microspores of rape (*Brassica napus* L.). *Plant Breeding* 100: 320–2.

McCaffrey J.P., Harmon B.L., Brown J., Brown A.P., Davis J.B. 1999. Assessment of *Sinapis alba*, *Brassica napus* and *S. alba* x *B. napus* hybrids for resistance to cabbage seedpod weevil, *Ceutorhynchus assimilis* (Coleoptera: Curculionidae). *J. Agr. Sci.* 132: 289–95.

McClellan D., Kott L.S., Beversdorf W.D., Ellis B.E. 1993. Glucosinolate metabolism in zygotic and microspore-derived embryos of *Brassica napus* L. *J. Plant Physiol.* 141: 153–9.

McClinchey S.L., Kott L.S. 2008. Production of mutants with high cold tolerance in spring canola (*Brassica napus*). *Euphytica* 162: 51–67.

Meng J., Shi S., Gan L., Li Z., Qu X. 1998. The production of yellow-seeded *Brassica napus* (AACC) through crossing interspecific hybrids of *B. campestris* (AA) and *B. carinata* (BBCC) with *B. napus*. *Euphytica* 103: 329–33.

Pechan P.M., Keller W.A. 1988. Identification of potentially embryogenic microspores in *Brassica napus*. *Physiol. Plant.* 74: 377–84.

Polsoni L., Kott L.S., Beversdorf W.D. 1988. Large-scale microspore culture technique for mutation-selection studies in *Brassica napus*. *Can. J. Bot.* 66: 1681–5.

Pomeroy M.K., Kramer J.K.G., Hunt D.J., Keller W.A. 1991. Fatty acid changes during development of zygotic and microspore derived embryos of *Brassica napus*. *Physiol. Plant.* 81: 447–54.

Rahman M.H. 2001. Production of yellow-seeded *Brassica napus* through interspecific crosses. *Plant Breeding* 120: 463–72.

Rajcan I., Kasha K.J., Kott L.S., Beversdorf W.D. 1999. Detection of molecular markers associated with linolenic and erucic acid levels in spring rapeseed (*Brassica napus* L.). *Euphytica* 105: 173–81.

Rashid A., Rakow G., Downey R.K. 1994. Development of yellow-seeded *Brassica napus* L. through interspecific crosses. *Plant Breeding* 112: 127–34.

Rashid A., Rakow G., Downey R.K. 1995. Agronomic performance and seed quality of black seeded cultivars and two sources of yellow seeded *Brassica napus*. In: *Proceeding of the 9th International Rapeseed Congress*. Cambridge, UK. pp. 1141–6.

Relf-Eckstein J., Rakow G., Raney J.P. 2003. Yellow-seeded *Brassica napus* – a new generation of high quality canola for Canada. In: *Proceeding of the 11th International Rapeseed Congress*. Copenhagen, Denmark. pp. 458–60.

Ripley V.L., Arnison P.G. 1990. Hybridization of *Sinapis alba* L. and *Brassica napus* L. via embryo rescue. *Plant Breeding* 104: 26–33.

Röbbelen G., Thies W. 1980. Biosynthesis of seed oil and breeding for improved oil quality of rapeseed. In: Tsunoda S., Hinata K., Gomez-Campo C. (eds) *Brassica Crops and Wild Allies*. Japn Sci Soc Press, Tokyo. pp. 253–83.

Roy N.N., Tarr A.W. 1985. IXLIN – An interspecific source for high linoleic and low linolenic acids in rapeseed (*Brasica napus* L.). *Z. Pflanzenzuchtung* 95: 201–9.

Schierholt A., Rücker B., Becker H.C. 2001. Inheritance of high oleic acid mutations in winter oilseed rape (*Brassica napus* L.). *Crop Sci.* 41: 1444–9.

Sharpe A.G., Lydiate D.J. 2003. Mapping the mosaic genotypes in a cultivar of oilseed rape (*Brassica napus*) selected via pedigree breeding. *Genome* 46: 461–8.

Shaw E.J., Fletcher R.S., Dosdall L.L., Kott L.S. 2009. Biochemical markers for cabbage seedpod weevil (*Ceutorynchus obstrictus* (Marsham)) resistance in canola (*Brassica napus* L.). *Euphytica* 170: 297–308.

Shirzadegan M. 1986. Inheritance of seed colour in *Brassica napus* L. *Z. Pflanzenzuchtung* 96: 140–6.

Siebel J., Pauls K.P. 1989. A comparison of anther and microspore culture as a breeding tool in *Brassica napus*. *Theor. Appl. Genet.* 78: 473–9.

Simbaya J., Slominski B.A., Rakow G., Campbell L.D., Downey R.K., et al. 1995. Quality characteristics of yellow-seeded *Brassica* seed meals: protein, carbohydrates, and dietary fiber components. *J. Agric. Food Chem.* 43: 2062–6.

Somers D.J., Friesen K.R.D., Rakow G. 1998. Identification of molecular markers associated with linoleic acid desaturation in *Brassica napus*. *Theor. Appl. Genet.* 96: 897–903.

Stefansson B.R., Kondra Z.P. 1975. Tower summer rape. *Can. J. Plant Sci.* 55: 343–4.

Swanson E.B., Coumans M.P., Wu S.C., Barsby T.L., Beversdorf W.D. 1987. Efficient isolation of microspores and the production of microspore-derived embryos from *Brassica napus*. *Plant Cell Rep.* 6: 94–7.

Swanson E.B., Coumans M.P., Brown G.L., Patel J.D., Beversdorf W.D. 1988. The characterization of herbicide tolerant plants in *Brassica napus* L. after *in vitro* selection of microspores and protoplasts. *Plant Cell Rep.* 7: 83–7.

Swanson E.B., Herrgesell M.J., Arnoldo M., Sippell D.W., Wong R.S.C. 1989. Microspore mutagenesis and selection: canola plants with field tolerance to the imidazolinones. *Theor. Appl. Genet.* 78: 525–30.

Tang Z.L., Li J.N., Zhang X.K., Chen L., Wang R. 1997. Genetic variation of yellow-seeded rapeseed lines (*Brassica napus* L.) from different genetic sources. *Plant Breeding* 116: 471–4.

Tanhuanpää P.K., Vilkki J.P., Vilkki H.J. 1995. Association of a RAPD marker with linolenic acid concentration in the seed oil of rapeseed (*Brassica napus* L.). *Genome* 38: 414–16.

Taylor D.C., Weber N. 1994. Microspore-derived embryos of *Brassicaceae* – model system for studies of storage lipid bioassembly and its regulation. *Eur. J. Lipid Sci. Technol.* 96: 228–35.

Thomas E., Wenzel G. 1975. Embryogenesis from microspores of *Brassica napus*. *Z. Pflanzenzuchtung* 74: 77–81.

Thomas P.M., Kondra Z.P. 1973. Maternal effects on the oleic, linoleic, and linolenic acid content of rapeseed oil. *Can. J. Plant Sci.* 53: 221–5.

Thormann C.E., Romero J., Mantet J., Osborn T.C. 1996. Mapping loci controlling the concentrations of erucic and linolenic acids in seed oil of *Brassica napus* L. *Theor. Appl. Genet.* 93: 282–6.

Uzunova M., Ecke W., Weissleder K., Röbbelen G. 1995. Mapping the genome of rapeseed (*Brassica napus* L.): Construction of an RFLP linkage map and localization of QTLs for seed glucosinolates content. *Theor. Appl. Genet.* 90: 194–204.

Van Deynze A.E., Beversdorf W.D., Pauls K.P. 1993. Temperature effects on seed color in black- and yellow-seeded rapeseed. *Can. J. Plant Sci.* 73: 383–7.

Van Deynze A.E., Pauls K.P. 1994. The inheritance of seed colour and vernalization requirement in *Brassica napus* using doubled haploid populations. *Euphytica* 74: 77–83.

Wiberg E., Rahlen L., Hellman M., Tillberg E., Glimelius K., et al. 1991. The microspore-derived embryo of *Brassica napus* L. as a tool for studying embryo-specific lipid biogenesis and regulation of oil quality. *Theor. Appl. Genet.* 82: 515–20.

Wong R.S.C., Swanson E. 1991. Genetic modification of canola oil: high oleic acid canola. In: Haberstroh C., Morris C.E. (eds) *Fat and Cholesterol Reduced Foods: Technologies and Strategies*. Portfolio Publishing Co., Texas, USA. pp. 153–64.

Zhao J., Meng J. 2003. Detection of loci controlling seed glucosinolates content and their association with *Sclerotinia* resistance in *Brassica napus*. *Plant Breeding* 122: 19–23.

11

Plant biodiversity and biotechnology

Naglaa A. Ashry,
Field Crops Research Institute, ARC, Egypt

DOI: 10.1533/9781908818478.205

Abstract: There are 20 000 to 25 000 protein-coding genes in typical crop plants like maize and soybean. The collection of traits displayed (phenotype) depends on the genes present in a genome (genotype). The appearance of specific traits depends on many factors, including whether the gene(s) responsible for the trait is/are expressed or non-expressed, the cell specificity of the expressed genes, and how the gene products interact with environmental factors. Molecular marker-assisted breeding (MAS) technology may be useful to survey heat tolerance in various genotypes, including landraces and wild relatives of cereals. In addition, the comparison of quantitative trait loci (QTLs) linked to stress tolerance in various cereals may help identify common loci or genes linked to drought and heat tolerance. The understanding and potential manipulation of the mechanisms of thermotolerance and response to stresses in cereals, either by transgenic approaches or by molecular breeding, will rely on further achievements in genomics, proteomics and metabolic profiling.

Key words: biodiversity, biotechnology, protein expression, protein networks, qualitative traits, photosynthesis, heat shock factor, heat shock proteins, functional genomics, proteomics, molecular markers, breeding, gene engineering.

11.1 Biodiversity

Biodiversity is a contraction of the term 'biological diversity', and refers to the diversity of 'life'. The multifaceted nature of the biodiversity continuum is recognized in the definition adopted by the 'Convention on Biological Diversity' 1992: 'the variability among living organisms from all sources including, inter alia, terrestrial, marine and other aquatic ecosystems and the ecological complexes of which they are a part; this includes diversity within species, between species and of ecosystems'.

11.1.1 Species diversity

Species are the central concept of biodiversity, and can be defined as populations of phenotypically similar organisms that routinely exchange genes under natural conditions, taking into consideration that the observed traits of an organism result from the interaction between the genetic makeup of the organism and its environment. At the species level, biodiversity conservation has frequently focused on endangered species, endemic species, or locales with a highly diverse complement of species.

11.1.2 Genetic diversity

Genetic diversity is a fundamental unit of biodiversity. The diversity of the genetic material of an organism is the underlying reason for the variability within and between species (Primak 1993). Again, genetic diversity can be viewed at three levels: diversity between individuals within one population; diversity between populations within one species; and diversity between different species. Genetic variability provides the resilience and adaptability, fitness and evolutionary flexibility that are required for the survival of organisms. Genetic fitness is a particularly important survival mechanism for the present rapidly changing environmental conditions.

11.2 Biotechnology

There are 20000 to 25000 protein-coding genes in typical crop plants like maize and soybean. The collection of traits displayed by any organism (phenotype) depends on the genes present in its genome (genotype). The appearance of any specific trait will also depend on many other factors, including whether the gene(s) responsible for the trait is/are expressed or non-expressed, the specific cells within which the genes are expressed, and how the genes and/or the gene products interact with environmental factors. Biotechnology is any technique that uses living organisms or parts thereof to make or modify a product or improve plants, animals, or microorganisms for specific uses. The primary biotechnology technique which concerns and directly affects biodiversity is that of genetic manipulation, which has a direct impact on biodiversity at the genetic level. By these manipulations, novel genes or gene fragments can be introduced into organisms (creating transgenics), or existing genes within an organism can be altered. Biotechnology techniques are initially applied to organisms within a laboratory environment (contained use). Once a potentially useful organism has been created in the laboratory, it is usually tested under the controlled conditions of a laboratory, growth chamber, or greenhouse. Products of biotechnology find application in the pharmaceutical and chemical industries and in agriculture (James 1999). At present, proof-of-concept experiments have demonstrated potential for insect resistance, nematode resistance, increased nutritional value, decreased human allergens, better post-harvest quality, new flower colors, and other traits (Auer 2011). There is a focus on new approaches based on small RNAs, RNA interference and production of RNA-mediated traits in plants (Auer 2011). Accordingly, new methods for risk analysis and for monitoring transgenes overcoming the pitfalls of polymerase chain reaction (PCR) methods are required to perform analyses of off-target effects and persistence of RNAs in the environment.

Independent studies on the effect of biotechnologically produced varieties are expected to provide neutral reports (*http://natureinstitute. org/nontarget/browse_newreports.php*) on the effects of trials on the biodiversity and the environment. In addition, there are agronomic practices that are better suited to third world countries, able to provide sustainable growth and performing better than plants obtained with the biotech approach (Holdrege 2012).

11.3 Heat stress tolerance in cereals

11.3.1 Effect of elevated temperature on plants

Changes in the global climate, notably in temporal temperature patterns, are predicted to have important consequences for crop production (Parry 1990; Barnabas et al. 2008). Both plant growth and development are affected by temperature (Porter and Moot 1998; Koyro et al. 2012). The most significant factors for heat stress-related yield loss in cereals include the high temperature-induced shortening of developmental phases, reduced light perception over the shortened life cycle, and disturbance of the processes associated with carbon assimilation (transpiration, photosynthesis, and respiration) (Stone 2001). Temperatures higher than 35°C significantly decrease the activity of ribulose 1,5-bisphosphate carboxylase/oxygenase (Rubisco), thereby limiting photosynthesis (Crafts-Brandner and Law 2000; Griffin et al. 2004). Higher plants exposed to excess heat, at least 5°C above their optimal growing conditions, exhibit a characteristic set of cellular and metabolic responses required for the plants to survive under the high-temperature conditions (Guy 1999; Koyro et al. 2012). These effects include a decrease in the synthesis of normal proteins and the accelerated transcription and translation of heat shock proteins (HSPs) (Bray et al. 2000), the production of phyto-hormones (ABA) and antioxidants (Maestri et al. 2002), and changes in the organization of cellular structures, including organelles and the cytoskeleton, and membrane functions (Weis and Berry 1988; Barnabas et al. 2008).

Heat stress during vegetative growth causes many physiological and metabolic changes, including alterations in hormone homeostasis. Some of the heat-induced processes at the cell, organ and whole-plant levels may be hormone regulated; others may be the consequence of a new hormonal status, altered by heat stress (Hoffmann and Parsons 1991; Maestri et al. 2002; Koyro et al. 2012). ABA is implicated in osmotic stress responses and mediates one of the intracellular dehydration signaling pathways (Davies and Jones 1991, Ahmadi and Baker 1999). In field conditions where water shortage and high-temperature stresses frequently occur simultaneously, ABA induction may also be an important component of thermo-tolerance (Fitter and Hay 1987; Gong et al. 1998; Al-Whaibi 2011).

Elevated temperatures may reduce the activities of antioxidant enzymes, as observed in maize (Gong et al. 1997; Wahid et al. 2007; Ahmad et al. 2010 a and b; Koyro et al. 2012). High temperature causes modifications in membrane functions mainly because of the alteration of membrane fluidity. In plant cells, membrane-based processes such as photosynthesis and respiration are especially important (Liu et al. 2012). The contribution of saturated lipids and protein components to membrane function under high-temperature stress needs further study. Heat stress results in the misfolding of newly synthesized proteins and the denaturation of existing ones. Protein thermo-stability is believed to be provided in part by chaperones, a specific class of proteins capable of assisting other proteins in proper post-translational folding and in maintaining them in a functional state (Ellis 1990; Lin et al. 2011;

Al-Whaibi 2011; Liu et al. 2012). Chaperone properties were identified as one of the major functions of HSPs. The involvement of several HSPs in the acquired thermo-tolerance of Arabidopsis is well demonstrated (Burke et al. 2000; Hong and Vierling 2000; Queitsch et al. 2000). Very little is known about the possible involvement of HSPs in the thermo-tolerance of cereals (Cooper and Ho 1983; Marmiroli et al. 1989; Lin et al. 2011). High-molecular-weight HSPs (HSP70, HSP90, HSP101) are characterized by a high level of sequence similarity within the plant kingdom. However, molecular diversity within families of high-molecular-weight HSPs suggests that even closely related members or alleles may vary in their specific functions. Furthermore, members of the same HSP family may function in different cell compartments (Boston et al. 1996; Maestri et al. 2002; Lin et al. 2011). HSP101 was shown to be a major component of thermo-tolerance in Arabidopsis (Hong and Vierling 2000; Queitsch et al. 2000). A homologous protein was isolated and cloned from wheat (Wells et al. 1998; Campbell et al. 2001). The heat tolerance of plants is a complex trait, most probably controlled by multiple genes. The comparative molecular biological analysis of heat-sensitive and heat-tolerant genotypes of Festuca (Zhang et al. 2005) revealed that heat-tolerant genotypes responded to stress by increasing the expression of genes participating in photosynthesis, protein synthesis and the preservation of cell status, and of those related to transcription factors. Heat stress transcription factors (HSFs) are the terminal components of signal transduction pathways mediating the activation of genes responsive to heat stress (Lin et al. 2011). The sequencing of the Arabidopsis genome revealed the unique complexity of the plant HSF family, with 21 members belonging to three classes and 14 groups (Nover et al. 2001). Of the 24 000 monitored genes of Arabidopsis, 11% showed a significant effect in the case of heat-stress treatment (Busch et al. 2005). Many of these proved to be under the control of HSF-1a or HSF-1b, as determined by the global transcriptional analysis of knockout mutants (Busch et al. 2005). HSF-1a/1b regulated genes were found to be involved in other functions, including protein biosynthesis and processing, signaling, metabolism, and transport. Furthermore, all the steps in the pathway resulting in osmolyte accumulation were reported to be HSF and/or heat regulated (Busch et al. 2005). It is documented that several steps in the heat response overlap with those involved in the response to various other forms of stress, such as drought and cold (Quinn 1988; Bray et al. 2000; Rizhsky et al. 2004; Swindell et al. 2007; Liu et al. 2012). Therefore, the investigation of the heat tolerance of the plants might provide useful information on general stress-tolerance mechanisms as well (Wahid et al.

2007; Li et al. 2009). The further understanding and potential manipulation of these mechanisms in cereals, either by transgenic approaches or by molecular breeding, will rely on further achievements in genomics, proteomics, and metabolic profiling (Zhang et al. 2004; Barnabas et al. 2008).

The effects of temperature on storage protein composition are unclear and may vary with genotype (Dupont and Altenbach 2003). Experimental results indicate that changes in the protein fraction composition under heat stress are mainly caused by the altered quantity of total N accumulated during grain filling (Triboï et al. 2003). In accordance, the post-anthesis application of N fertilizers reduced the effect of high temperature on the storage protein composition of wheat (Dupont and Altenbach 2003).

11.4 Modern approaches in cereals for yield and food security under temperature stress

Environmental stresses have a great impact on the yield of cereal crops. As detailed earlier, the effect of heat stress on yield is highly complex and involves processes as diverse as stem reserve accumulation, gametogenesis, fertilization, embryogenesis, and endosperm and grain development. Our present knowledge on these processes and on their mutual interactions is still scant, especially if the potential impacts of environmental factors also have to be considered. The application of modern research tools to reveal the complex molecular networks behind the observed physiological and developmental responses in higher plants, including cereals, has only recently begun (Snape et al. 2005; Langridge et al. 2006; Varshney et al. 2006; Sreenivasulu et al. 2007; Al-Whaibi 2011).Therefore, only a few examples directly related to biodiversity assessment, identification, and preservation are discussed further in detail.

11.4.1 Use of molecular markers and breeding

Genetic maps based on molecular marker technologies are now available for major cereal species (Snape et al. 2005; Langridge et al. 2006; Phumichai et al. 2008; Koyro et al. 2012). Many of the traits determining abiotic stress tolerance (Park et al. 2007) and the quality and quantity of yield are controlled by a large number of genes, which have only minor individual effects but which act together (quantitative trait loci, QTL). In crop species with large, complex genomes, QTL analysis is an important tool in the identification of genetic markers to assist breeding efforts. This approach is complicated in wheat because of the polyploid nature of the genome and the low levels of polymorphism, but is straightforward in rice, maize, and barley (Snape et al. 2005; Huang et al. 2009). The strong resemblance observed between the genetic maps of cereals, however, may help transfer the knowledge gained for rice or barley to wheat. Studies on the abiotic stress tolerance of cereals include the extensive analysis of QTLs linked to the field evaluation of stress tolerance (Langridge et al. 2006). The genetic bases of traits representing source, sink, and transport tissues and their relationship to yield have been investigated by QTL analysis in rice (Cui et al. 2003; Abdelkhalik et al. 2005; Wahid et al. 2007; Vasquez-Robinet et al. 2010). Correlating genetic information with physiological and morphological traits related to high yield and/or heat tolerance will allow the development of new varieties with improved yield safety under heat stress conditions using molecular marker-assisted breeding (MAS). This technology may also be useful to survey heat tolerance in various genotypes, including landraces and wild relatives of cereals. In addition, the comparison of QTLs linked to stress tolerance in various cereals may help identify common loci or genes linked, for example, to drought tolerance (Langridge et al. 2006) and heat tolerance (Park et al. 2007; Vasquez-Robinet 2010; Huang et al. 2012).

Among the cereals, transcription of retrotransposons appears to be a common phenomenon, corresponding to about 0.1% of total transcripts in expressed sequence tag (EST) databases (Schulman et al. 2004). Stress appears to be a general phenomenon, which has been well analyzed in tobacco and rice (Beguiristain et al. 2001). In barley, although stress activation has not been directly demonstrated, both retrotransposon BARE-1 copy number and genome size are correlated with environmental factors associated with drought and temperature stress (Kalendar et al. 2000).

The increase in temperatures and climate changes are causing decreases in crop yields, especially in the African regions with tropical and semi-arid climate. For this reason a breeding program was started to produce inbred lines of maize possessing characters from heat-resistant varieties able to cope with heat stress. The Quality Protein Maize (QPM) varieties based on the Opapue-2 (a bZip transcription regulatory factor) mutation by the International Maize and Wheat Improvement Center (CIMMYT) were found to contain about twice the levels of lysine and tryptophan and 10% higher grain yield than the most modern varieties of tropical maize. High levels of these two amino acids enhance manufacture of complete proteins in the body. QPM has similar qualities to normal maize in grain texture, taste, color, tolerance to biotic and abiotic stresses as well as high yield, and performs like normal maize (Sofi et al. 2009; Olaoye et al. 2009). Since 1993, 33 tropical and 22 subtropical QPM lines have been released as CIMMYT Maize Lines (CMLs).

The major grain crops, maize, wheat and rice, do not have commercial perennial varieties. However, all have wild relatives that are perennials, and advances in genetics and breeding enable relatively rapid breeding programs that were not possible a decade ago. The primary strategy is to introgress genes for perennialism from wild relatives into current grain crops, via either accelerated breeding and/or genetic modification (GM), and evaluate the potential of non-domesticated perennials for development as crops. There are small-scale efforts underway across the globe, but there is no large-scale, coordinated effort to develop perennialism.

Maize lines need to be genetically characterized, especially for phenotyping the heat-responsive activation of hormone biosynthesis pathways, and genes involved in adaptation to high temperatures. Structural genes of the jasmonate (JA) biosynthetic pathway, namely lipoxygenase, allene oxide synthase, allene oxide cyclase, and oxo-phytodienoate reductase, directly involved in the biosynthesis of JA in plant roots, may behave differently in stress-tolerant and non-tolerant varieties of maize, as has been shown in other crops. Regulatory genes of JA biosynthesis, that is, MYC2 transcription factors, and the signaling molecule moving through the phloem to the roots, miR-319, targeting transcription factors of the TCP family that regulate several genes of the lipoxygenase pathway (Schommer et al. 2008), need to be monitored for differential expression.

11.4.2 Functional genomics

Expression profiling may help to identify the key molecular events underlying stress tolerance and grain development, as well as their interactions. The number of cereal EST sequences available in public databases is continuously increasing (Pavli et al. 2011). The cDNA libraries used to generate these ESTs represent various tissues and growth conditions (Alexandrov et al. 2009, Koyro et al. 2012), but yield- and stress-related libraries dominate. ESTs are especially important in wheat genomics, because of the size and complexity of the genome (Langridge et al. 2006; Houde et al. 2006; Timperio et al. 2008). These authors reported that the digital expression analysis of EST sequences combined with gene annotation (annotation of 29 556 different sequences) resulted in the identification of several pathways associated with abiotic stress resistance in wheat.

Modern molecular techniques have made it possible to isolate mRNA from a single cell; this facilitates the study of gene expression patterns in isolated egg cells or zygotes using PCR (Brandt et al. 2002; Okamoto et al. 2005; Lin et al. 2011; Liu et al. 2012), genomic methods (Brandt et al. 2002; Sprunck et al. 2005; Al-Whaibi 2011) and proteomic methods (Okamoto et al. 2004; Wang et al. 2009; Vasquez-Robinet et al. 2010; Neilson et al. 2010; Miernyk and Hajduch 2011; Huang et al. 2012). It is expected that these new techniques will accelerate the accumulation of acquired knowledge about the mechanisms by which plants tolerate abiotic stresses.

11.4.3 Proteomics, metabolomics

Investigating the effect of heat stress on protein composition might also be an important step towards understanding the link between environmental factors and plant development. Proteomic studies in cereals can be based on rice as a model species (Agrawal and Rakwal 2005; Agrawal et al. 2006; Komatsu and Yano 2006). Proteomic analysis has been started in all major cereal species in addition to rice (Wang et al. 2009). Proteomic reference maps have been compiled for maize (Méchin et al. 2004) and wheat (Vensel et al. 2005) endosperm and for barley grain (Finnie et al. 2002) during the processes of grain filling and maturation. The effect of heat stress on the grain of hexaploid wheat has been thoroughly studied at the protein level (Majoul et al. 2003, 2004; Wahid et al. 2007; Li et al. 2009; Vasquez-Robinet et al. 2010; Liu et al. 2012; Huang et al. 2012). The down-regulation of several proteins involved in starch metabolism and the induction of HSPs were reported (Majoul et al. 2003; Timperio et al. 2008; Yu et al. 2012). As regards the effect of drought on the wheat grain proteome, Hajheidari et al. (2007) reported the detection of 121 proteins exhibiting significant changes in response to the stress, of which 57 could be identified. Two-thirds of the identified proteins turned out to be thioredoxin targets, revealing the link between drought and oxidative stresses. Metabolomic research on cereals has also recently begun (Sato et al. 2004) and may, in the future, provide valuable information, for instance, on the sugar and amino acid metabolism in the vegetative and reproductive organs of cereals under various environmental conditions (Langridge et al. 2006; Wang et al. 2009; Neilson et al. 2010; Miernyk and Hajduch 2011; Huang et al. 2012).

11.4.4 Genetic engineering

Genetic modification allows the introduction of isolated individual genes into the cereal germplasm and offers a variety of opportunities to increase environmental stress tolerance. The major cereal species are all amenable to the technologies of genetic modification (Shrawat and Lörz 2006; Vasil 2007). Although there is still a political debate on the intensive use of genetically modified plants in agriculture, especially in Europe, there is a continuous increase in transgenic crop production worldwide, including many varieties of genetically modified maize resistant to herbicide and/or pathogens. As knowledge on the molecular networks underlying abiotic stress responses in plants, especially in Arabidopsis (Zhu et al. 2008), continuously increases (Zhang et al. 2004; Nakashima and Yamaguchi-Shinozaki 2005), more and more candidate genes will be identified with the potential to improve stress tolerance using transgenesis. In addition to model species such as Arabidopsis, several candidate genes have also been identified in cereals themselves (Van De Wiel and Lotz 2006) through the application of modern genomic approaches (Shinozaki and Yamaguchi-Shinozaki 2007). Further, strategies used to successfully improve drought or heat tolerance in transgenic cereals included the overexpression of kinases or phosphatases involved in the stress signaling cascade (Saijo et al. 2000; Shou et al. 2004; Xu et al. 2007), osmoprotectants (Garg et al. 2002; Abebe et al. 2003; Jang et al. 2003; Capell et al. 2004; Quan et al. 2004; Shirasawa et al. 2006) or other stress-related genes (Shi et al. 2001; Katiyar-Agarwal et al. 2003). Most of these transgenic strategies have an indirect effect on reproductive development by ensuring normal plant growth and metabolism under adverse environmental conditions. However, as discussed earlier, photosynthesis, CO_2 assimilation, carbon and nitrogen metabolism, and nutrient transport during vegetative and reproductive development all have very important roles in the development of the reproductive organs under both normal and stress conditions. In addition to these general stress-resistance traits, which might also be successfully used to ensure high yield under water limitation and/or temperature stress, more and more genetic modification of specific pathways associated with reproduction and seed development can be expected in the future.

11.5 Future perspectives

The complexity of both cereal reproduction and plant stress responses makes it difficult to construct a simple model of ways in which successful reproductive development and high yield can be achieved under water-limited and/or high-temperature conditions. Both specific and more general approaches are conceivable, targeting various aspects of plant development and stress responses. However, since the safety of the final yield is the main target, all the breeding or genetic manipulation approaches used in cereals have to converge finally at flowering and/or grain development. Therefore, the better understanding of these developmental processes is of utmost importance for the future, and modern genomic approaches may help considerably in this respect. However, cereal yield is not only dependent on the success of the reproductive processes themselves, but is indirectly determined by overall plant growth and development as well. This is well supported by the breeding of modern cultivars with reduced plant height (most of the cereal species) or tassel size (maize).

Finally, it is unfair to generally state that there is an inverse relationship between biodiversity and biotechnology, and that biotechnology has had only bad effects on biodiversity; rather, every technology has two sides, and wise and well-monitored usage of any technology is the sole assurance of avoiding harmful effects. One cannot deny that the broad approach of biotechnology could be the key to solving, or at least paving the road towards solving, the world's food and feed problems, but only if it is used in the correct way. It is still the responsibility of scientists to show, transparently, both sides of a specific technique for decision-makers in the best interests of humanity.

11.6 Acknowledgment

The author wishes to express her deep gratitude to the members of 'Ibn Sina' Group, an Egyptian scientific web-based network, for their unlimited support to fulfill this work.

11.7 References

Abdelkhalik A.F., Shishido R., Nomura K. and Ikehashi H. (2005) QTL-based analysis of leaf senescence in an indica/japonica hybrid in rice (*Oryza sativa* L.). *Theoretical and Applied Genetics* 110, 1226–35.

Abebe T., Guenzi A.C., Martin B. and Cushman J.C. (2003) Tolerance of mannitol-accumulating transgenic wheat to water stress and salinity. *Plant Physiology* 131, 1748–55.

Agrawal G.K. and Rakwal R. (2005) Rice proteomics: a cornerstone for cereal food crop proteomes. *Mass Spectrometry Reviews* 25, 1–53.

Agrawal G.K., Jwa N., Iwahashi Y., Yonekura M., Iwahashi H. et al. (2006) Rejuvenating rice proteomics: facts, challenges, and visions. *Proteomics* 6, 5549–76.

Ahmad P., Jaleel C.A. and Sharma S. (2010a) Antioxidative defence system, lipid peroxidation, proline metabolizing enzymes and biochemical activity in two genotypes of Morus *alba* L. subjected to NaCl stress. *Russian Journal of Plant Physiology* 57, 509–17.

Ahmad P., Jaleel C.A., Salem M.A., Nabi G. and Sharma S. (2010b) Roles of enzymatic and non-enzymatic anti-oxidants in plants during abiotic stress. *Critical Reviews in Biotechnology* 30, 161–75.

Ahmadi A. and Baker D.A. (1999) Effect of abscisic acid (ABA) on grain filling processes in wheat. *Plant Growth Regulation* 28, 187–97.

Alexandrov N.N., Brover V.V., Freidin S., Troukhan M.E., Tatarinova T.V., et al. (2009) Insights into corn genes derived from large-scale cDNA sequencing. *Plant Mol Biol.* 69, 79–94.

Al-Whaibi M.H. (2011) Plant heat-shock proteins: A mini review. *Journal of King Saud University*, 23, 139–50.

Auer C. (2011) Small RNAs for crop improvement. In Erdmann V.A., Barciszewski J. (Eds) *Non Coding RNAs in Plants*. Springer-Verlag, Berlin. pp. 461–84.

Barnabas B., Jager K. and Feher A. (2008) The effect of drought and heat stress on reproductive processes in cereals. *Plant Cell Environment* 31, 11–38.

Beguiristain T., Grandbastien M.A., Puigdomènech P. and Casacuberta J.M. (2001) Three Tnt1 subfamilies show different stress-associated patterns of expression in tobacco. Consequences for retrotransposon control and evolution in plants. *Plant Physiology* 127, 212–21.

Boston R.S., Viitanen P.V. and Vierling E. (1996) Molecular chaperones and protein folding in plants. *Plant Molecular Biology* 32, 191–222.

Brandt S., Kloska S., Altmann T. and Kehr J. (2002) Using array hybridization to monitor gene expression at the single cell level. *Journal of Experimental Botany* 53, 2315–23.

Bray E.A., Bailey-Serres J. and Weretilnyk E. (2000) Responses to abiotic stresses. In B. Buchanan, W. Gruissem and R. Jones (Eds) *Biochemistry and Molecular Biology of Plants*. Rockville, MD, USA, ASPB. pp. 1158–203.

Burke J.J., O'Mahony P.J. and Oliver M.J. (2000) Isolation of Arabidopsis mutants lacking components of acquired thermotolerance. *Plant Physiology* 123, 575–87.

Busch W., Wunderlich M. and Schoffl F. (2005) Identification of novel heat shock factor-dependent genes and biochemical pathways in *Arabidopsis thaliana*. *The Plant Journal* 41, 1–14.

Campbell J.L., Klueva N.Y., Zheng H., Nieto-Sotelo J., Jo T.H.D. et al. (2001) Cloning of new members of heat shock protein HSP101 gene family in wheat (*Triticum aestivum* (L.) Moench) inducible by heat, dehydration, and ABA. *Biochimica and Biophysica Acta* 1517, 270–7.

Capell T., Bassie L. and Christou P. (2004) Modulation of the polyamine biosynthetic pathway in transgenic rice confers tolerance to drought stress. *Proceedings of the National Academy of Sciences of the U S A* 101, 9909–14.

Cooper P. and Ho T. (1983) Heat shock proteins in maize. *Plant Physiology* 71, 215–22.

Crafts-Brandner S.J. and Law R.D. (2000) Effect of heat stress on the inhibition and the recovery of the ribulose-1,5-bisphosphate carboxylase/oxygenase activation state. *Planta* 212, 67–74.

Cui K.H., Peng S.B., Xing Y.Z., Yu S.B., Xu C.G. et al. (2003) Molecular dissection of the genetic relationships of source, sink and transport tissue with yield traits in rice. *Theoretical and Applied Genetics* 106, 649–58.

Davies W.J. and Jones H.G. (1991) *Abscisic Acid: Physiology and Biochemistry*. BIOS Scientific Publishers, Oxford, UK.

Dupont F. and Altenbach S. (2003) Molecular and biochemical impacts of environmental factors on wheat grain development and protein synthesis. *Journal of Cereal Science* 38, 133–46.

Ellis R.J. (1990) The molecular chaperone concept. *Seminars in Cell Biology* 1, 1–9.

Finnie C., Melchior S., Roepstorff P. and Svensson B. (2002) Proteome analysis of grain filling and seed maturation in barley. *Plant Physiology* 129(3), 1308–19.

Fitter A.H. and Hay R.K.M. (1987) *Environmental Physiology of Plants*. Academic Press, London, UK.

Garg A.K., Kim J., Owens T.G., Ranwala A.P., Choi Y.D., et al. (2002) Trehalose accumulation in rice plants confers high tolerance levels to different abiotic stresses. *Proceedings of the National Academy of Sciences of the USA* 99, 15 898–903.

Gong M., Chen S.N., Song Y.Q. and Li Z.G. (1997). Effect of calcium and calmodulin on intrinsic heat tolerance in relation to antioxidant systems in maize seedlings. *Australian Journal of Plant Physiology* 24, 371–9.

Gong M., Li Y.J. and Chen S.Z. (1998) Abscisic acid-induced thermotolerance in maize seedlings is mediated by calcium and associated with antioxidant systems. *Journal of Plant Physiology* 153, 488–96.

Griffin J.J., Ranney T.G. and Pharr D.M. (2004) Heat and drought influence photosynthesis and water relations, and soluble carbohydrates of two ecotypes of redbud (*Cercis canadensis*). *Journal of American Society for Horticultural Science* 129, 497–502.

Guy C. (1999). The influence of temperature extremes on gene expression, genomic structure, and the evolution of induced tolerance in plants. In H.R. Lerner (Ed.) *Plant Responses to Environmental Stresses*. Marcel Dekker, New York, NY. pp. 497–548.

Hajheidari M., Eivazi A., Buchanan B.B., Wong J.H., Majidi I. et al. (2007) Proteomics uncovers a role for redox in drought tolerance in wheat. *Journal of Proteome Research* 6, 1451–60.

Hoffmann A.A. and Parsons P.A. (1991) *Evolutionary Genetics and Environmental Stress*. Oxford University Press, Oxford, UK.

Holdrege K. (2012) Context-sensitive action: the development of push-pull farming in Africa. *In Context*, 27, 11–16, The Nature Institute, Canada.

Hong S.W. and Vierling E. (2000) Mutants of *Arabidopsis thaliana* defective in the acquisition of tolerance to high temperature stress. *Proceedings of the National Academy of Sciences of the USA* 97, 4392–7.

Houde M., Belcaid M., Ouellet F., Danyluk J., Monroy A.F., et al. (2006) Wheat EST resources for functional genomics of abiotic stress. *BMC Genomics* 7, 149.

Huang H., Miller I.M. and Song S.-Q. (2012) Proteomics of desiccation tolerance during development and germination of maize embryos. *Journal of Proteomics* 75, 1247–62.

Huang X., Liu L., Y. Zhai, Liu T. and Chen J. (2009) Proteomic comparison of four maize inbred lines with different levels of resistance to *Curvularia lunata* (Wakker) Boed infection. *Progress in Natural Science* 19, 353–8.

James, C. 1999. *Review of Commercialised Transgenic Crops*. ISAAA Briefs No. 12, ISAAA, Ithaca, NY.

Jang I.C., Oh S., Seo J., Choi WB, Song SI, et al. (2003) Expression of a bifunctional fusion of the *Escherichia coli* genes for trehalose-6-phosphate synthase and trehalose-6-phosphate phosphatase in transgenic rice plants increases trehalose accumulation and abiotic stress tolerance without stunting growth. *Plant Physiology* 131, 516–24.

Kalendar R., Tanskanen J., Immonen S., Nevo E. and Schulman A.H. (2000) Genome evolution of wild barley (*Hordeum spontaneum*) by BARE-1 retrotransposon dynamics in response to sharp microclimatic divergence. *Proceedings of the National Academy of Sciences of the USA* 97, 6603–7.

Katiyar-Agarwal S., Agarwal M. and Grover A. (2003) Heat-tolerant basmati rice engineered by overexpression of hsp101. *Plant Molecular Biology* 51, 677–86.

Komatsu S. and Yano H. (2006) Update and challenges on proteomics in rice. *Proteomics*, 6, 4057–68.

Koyro H.-W., Ahmad P. and Geissler N. (2012). Abiotic stress responses in plants: An overview. In Ahmad P. and Prasad M.N.V. (Eds) *Environmental Adaptations and Stress Tolerance of Plants in the Era of Climate Change*. Springer, NY.

Langridge P., Paltridge N. and Fincher G. (2006) Functional genomics of abiotic stress tolerance in cereals. *Briefings of Functional Genomics and Proteomics* 4, 343–54.

Li H.-Y., Huang S.-H., Shi Y.-S., Song Y.-C., Zhong Z.-B., et al. (2009) Isolation and analysis of drought-induced genes in maize roots. *Agricultural Sciences in China*, 8(2), 129–36.

Lin Y.-X., H.-Y. Jiang, Chu Z.-X., Tang X.-L., Zhu S.-W., et al. (2011) Genome-wide identification, classification and analysis of heat shock transcription factor family in maize. *BMC Genomics* 12, 76.

Liu Y.-J., Yuan Y., Liu Y.-Y., Liu Y., Fu J.-J., et al. (2012) Gene families of maize glutathione–ascorbate redox cycle respond differently to abiotic stresses. *Journal of Plant Physiology* 169, 183–92.

Maestri E., Klueva N., Perrotta C., Gulli M., Nguyen T. et al. (2002) Molecular genetics of heat tolerance and heat shock proteins in cereals. *Journal of Plant Molecular Biology* 48, 667–81.

Majoul T., Bancel E., Triboï E., Ben Hamida J. and Branlard G. (2003) Proteomic analysis of the effect of heat stress on hexaploid wheat grain: characterization of heat-responsive proteins from total endosperm. *Proteomics* 3, 175–83.

Majoul T., Bancel E., Triboï E., Ben Hamida J. and Branlard G (2004) Proteomic analysis of the effect of heat stress on hexaploid wheat grain: characterization of heat-responsive proteins from non-prolamins fraction. *Proteomics* 4, 505–13.

Marmiroli N., Lorenzoni C, Cattivelli L., Stanca A.M. and Terzi V. (1989) Induction of heat shock proteins and acquisition of thermotolerance in barley (*Hordeum vulgare* L.): variations associated with growth habit and plant development. *Journal of Plant Physiology* 135, 267–73.

Méchin V., Balliau T., Château-Joubert S., Davanture M., Langella O., et al. (2004). A two-dimensional proteome map of maize endosperm. *Phytochemistry* 65, 1609–18.

Miernyk J.A. and Hajduch M. (2011) Seed proteomics. *J. Proteomics*, 74, 389–400.

Nakashima K. and Yamaguchi-Shinozaki K. (2005) Molecular studies on stress-responsive gene expression in Arabidopsis and improvement of stress tolerance in crop plants by regulon biotechnology. *Japan Agricultural Research Quarterly* 39, 221–9.

Neilson K.A., Gammulla C.G., Mirzaei M., Imin N. and Haynes P.A. (2010) Proteomic analysis of temperature stress in plants. *Proteomics* 10, 828–45.

Nover L., Bharti K., Doring P., Mishra S.K., Ganguli A. et al. (2001) Arabidopsis and the heat stress transcription factor world: how many heat stress transcription factors do we need? *Cell Stress and Chaperones* 6, 177–89.

Okamoto T., Higuchi K., Shinkawa T., Isobe T., Lorz H., et al. (2004) Identification of major proteins in maize egg cells. *Plant and Cell Physiology* 45, 1406–12.

Okamoto T., Scholten S., Lorz H. and Kranz E. (2005) Identification of genes that are up- or down-regulated in the apical or basal cell of maize two-celled embryos and monitoring their expression during zygote development by a cell manipulation- and PCR-based approach. *Plant and Cell Physiology* 46, 332–8.

Olaoye G., Bello O.B., Ajani A.K. and Ademuwagun T.K. (2009) Breeding for improved organoleptic and nutritionally acceptable green maize varieties by crossing sweet corn (*Zea mays saccharata*): Changes in quantitative and qualitative characteristics in F_1 hybrids and F_2 populations. *Journal of Plant Breeding and Crop Science* 1(9), 298–305.

Park S.-C., Lee J.R., Shin S.-O., Park Y., Lee S.Y., et al. (2007) Characterization of a heat-stable protein with antimicrobial activity from *Arabidopsis thaliana*. *Biochemical and Biophysical Research Communications* 362, 562–7.

Parry M.L. (1990) *Climate Change and World Agriculture*. Earthscan Publications, London.

Pavli O.I., Ghikas D.V., Katsiotis A. and Skaracis G.N. (2011) Differential expression of heat shock protein genes in sorghum (*Sorghum bicolor* L.) genotypes under heat stress. *Australian Journal of Crop Science* 5(5), 511–15.

Phumichai C., Doungchan W., Puddhanon P., Jampatong S., Grudloyma P., et al. (2008) SSR-based and grain yield-based diversity of hybrid maize in Thailand. *Field Crops Research* 108, 157–62.

Porter J.R. and Moot D.J. (1998) Research beyond the means: climatic variability and plant growth. In Dalezios N.R. (Ed.) *International Symposium on Applied Agro-meteorology and Agro-climatology*. Office for Official Publication of the European Commission, Luxembourg. pp. 13–23.

Primak, R.B. (1993) *Essentials of Conservation Biology*. Sinauer Associates Inc., Sunderland, Massachusetts, USA.

Quan R., Shang M., Zhang H., Zhao Y. and Zhang J. (2004) Engineering of enhanced glycine betaine synthesis improves drought tolerance in maize. *Plant Biotechnology Journal* 2, 477–86.

Queitsch C., Hong S.W., Vierling E. and Lindquist S.L. (2000) Heat shock protein 101 plays a crucial role in thermo-tolerance in Arabidopsis. *Plant Cell* 12, 479–92.

Quinn P.J. (1988) Effects of temperature on cell membranes. *Symposia of the Society for Experimental Biology* 42, 237–58.

Rizhsky L., Liang H., Shuman J., Shulaev V., Davletova S. et al. (2004) When defense pathways collide: the response of Arabidopsis to a combination of drought and heat stress. *Plant Physiology* 134, 1683–96.

Saijo Y., Hata S., Kyozuka J., Shimamoto K. and Izui K. (2000) Over-expression of a single Ca_2+-dependent protein kinase confers both cold and salt/drought tolerance on rice plants. *Plant Journal* 23, 319–27.

Sato S., Soga T., Nishioka T. and Tomita M. (2004) Simultaneous determination of the main metabolites in rice leaves using capillary electrophoresis mass spectrometry and capillary electrophoresis diode array detection. *Plant Journal* 40, 151–63.

Schommer C., Palatnik J., Aggarwal P., Chetelat A., Cubas P., et al. (2008) Control of jasmonate biosynthesis and senescence by miR319 targets. *PLoS Biology* 6, e230.

Schulman A.H., Gupta P.K. and Varshney R.K. (2004) Organization of retrotransposons and microsatellites in cereal genomes. In P.K. Gupta and R.K. Varshney (Eds) *Cereal Genomics*. Kluwer, Amsterdam. pp. 83–118.

Shi W.M., Muramoto Y., Ueda A. and Takabe T. (2001) Cloning of peroxisomal ascorbate peroxidase gene from barley and enhanced thermo-tolerance by over-expressing in *Arabidopsis thaliana*. *Gene* 273, 23–7.

Shinozaki K. and Yamaguchi-Shinozaki K. (2007) Gene networks involved in drought stress response and tolerance. *Journal of Experimental Botany* 58, 221–7.

Shirasawa K., Takabe T., Takabe T. and Kishitani S. (2006) Accumulation of glycinebetaine in rice plants that overexpress choline monooxygenase from spinach and evaluation of their tolerance to abiotic stress. *Annals of Botany* 98, 565–71.

Shou H., Bordallo P. and Wang K. (2004) Expression of the *Nicotiana* protein kinase (NPK1) enhanced drought tolerance in transgenic maize. *Journal of Experimental Botany* 55, 1013–19.

Shrawat A.K. and Lörz H. (2006) Agrobacterium-mediated transformation of cereals: a promising approach crossing barriers. *Plant Biotechnology Journal* 4, 575–603.

Snape J., Fish L., Leader D., Bradburne R. and Turner A. (2005) The impact of genomics and genetics on wheat quality improvement. *Turkish Journal of Agriculture and Forestry* 29, 97–103.

Sofi P.A., Shafiq A., Wani A., Rather G. and Shabir H.W. (2009) Quality protein maize (QPM): genetic manipulation for the nutritional fortification of maize. *Journal of Plant Breeding and Crop Science* 1, 244–53.

Sprunck S., Baumann U., Erwards K. and Langridge P. (2005) The transcript composition of egg cells changes significantly following fertilization in wheat (*Triticum aestivum* L.). *Plant Journal* 41, 660–72.

Sreenivasulu N., Sopory S.K. and Kavi Kishor P.B. (2007) Deciphering the regulatory mechanisms of abiotic stress tolerance in plants by genomic approaches. *Gene* 388, 1–13.

Stone P. (2001) The effects of heat stress on cereal yield and quality. In Basra A.S. (Ed.) *Crop Responses and Adaptations to Temperature Stress*. Food Products Press, Binghamton, NY. pp. 243–91.

Swindell W.R., Huebner M. and Weber A.P. (2007) Transcriptional profiling of Arabidopsis thaliana heat shock proteins and transcription factors reveals extensive overlap between heat and non-heat stress response pathways. *BMC Genomics* 8, 125.

Timperio A.M., Egidi M.G. and Zolla L. (2008) Proteomics applied on plant abiotic stresses: Role of heat shock proteins (HSP). *Journal of Proteomics* 71, 391–411.

Triboï E., Martre P. and Triboï-Blondel A.M. (2003) Environmentally-induced changes of protein composition for developing grains of wheat are related to changes in total protein content. *Journal of Experimental Botany* 54, 1731–42.

Van De Wiel C.C.M and Lotz L.A.P. (2006) Outcrossing and coexistence of genetically modified with (genetically) unmodified crops: a case study of the situation in the Netherlands. *NJAS-Wageningen Journal of Life Sciences* 54, 17–35.

Varshney R.K., Hoisington D.A. and Tyagi A.K. (2006) Advances in cereal genomics and applications in crop breeding. *Trends in Biotechnology* 24, 490–9.

Vasil I.K. (2007) Molecular genetic improvement of cereals: transgenic wheat (*Triticum aestivum* L.). *Plant Cell Reports* 26, 1133–54.

Vasquez-Robinet C., Watkinson J.I., Allan A.S., Naren R., Lenwood S.H. et al. (2010) Differential expression of heatshock protein genes in preconditioning for photosynthetic acclimation in water-stressed loblolly pine. *Plant Physiology and Biochemistry* 48, 256–64.

Vensel W.H., Tanaka C.K., Cai N., Wong J.H., Buchanan B.B. et al. (2005) Developmental changes in the metabolic protein profiles of wheat endosperm. *Proteomics* 5, 1594–611.

Wahid A., Gelani S., Ashraf M. and Foolad M.R. (2007) Heat tolerance in plants: An overview. *Environmental and Experimental Botany* 61, 199–223.

Wang J.W., Yang F.P., Chen X.Q., Liang R.Q., Zhang L.Q., et al. (2009) Induced expression of *dreb* transcriptional factor and study on its physiological effects of drought tolerance in transgenic wheat. *Acta Genetica Sinica* 33, 468–76.

Weis E. and Berry J.A. (1988) Plants and high temperature stress. In Long S.P. and Woodward F.I. (Eds) *Plants and Temperature*. Company of Biologists Ltd, Cambridge, UK. pp. 329–46.

Wells D.R., Tanguay R.L., Le H. and Gallie D.R. (1998) HSP101 functions as a specific translational regulatory protein whose activity is regulated by nutrient status. *Genes and Development* 12, 3236–51.

Xu C., Jing R., Mao X., Jia X. and Chang X. (2007) Wheat (*Triticum aestivum*) protein phosphatase 2a catalytic subunit gene provides enhanced drought tolerance in tobacco. *Annals of Botany* 99, 439–50.

Yu Y.L., Zhang H., Li W., Mu C., Zhang F., et al. (2012) Genome-wide analysis and environmental response profiling of the FK506-binding protein gene family in maize (*Zea mays* L.). *Gene* 498, 212–22.

Zhang J.Z., Creelman R.A. and Zhu J.K. (2004) From laboratory to field. Using information from Arabidopsis to engineer salt, cold, and drought tolerance in crops. *Plant Physiology* 135, 615–21.

Zhang Y., Mar M., Chekhovsky K., Kupfer D., Lai H. et al. (2005) Differential gene expression in Festuca under heat stress conditions. *Journal of Experimental Botany* 56, 897–907.

Zhu B., Xiong A.-S., Peng R.-H., Xu J., Zhou J., et al. (2008) Heat stress protection in Aspen sp1 transgenic *Arabidopsis thaliana*. *BMB Reports* 31, 382–7.

Natural resveratrol bioproduction

Angelo Santino, Marco Taurino, Ilaria Ingrosso and Giovanna Giovinazzo, Institute of Sciences of Food Productions, CNR-ISPA, Italy

DOI: 10.1533/9781908818478.223

Abstract: Polyphenols form a large family of plant secondary metabolites. Since they are abundant in fruits and vegetables, they form an integral part of the human diet. Diverse health-related effects are attributed to flavonoids, including scavenging of free radicals and modification of enzymatic activities. Consequently, the ingestion of different flavonoid species may be more beneficial for human health than the ingestion of only a few species. However, in most fruit and crop plants, only one or a few flavonoid classes are present. One group of polyphenols believed to have a positive effect on health is the flavonoid-related class of stilbenes, specifically *trans*-resveratrol and its glucoside piccid. *trans* Resveratrol, mainly found in grape, peanut and a few other plants, displays a wide range of biological effects. There is growing interest in the development of plant cell cultures and food crops with tailor-made levels and composition of stilbenes, designed for optimal biological effect.

Key words: resveratrol, plant biotechnology, bioproduction.

12.1 Stilbenes and resveratrol

Chalcone synthase, the enzyme responsible for the first step in the flavonoid pathway (Figure 12.1), is a member of the plant polyketide synthase superfamily, which catalyses the production of a wide variety of secondary metabolites from a limited set of substrates (malonyl-CoA and

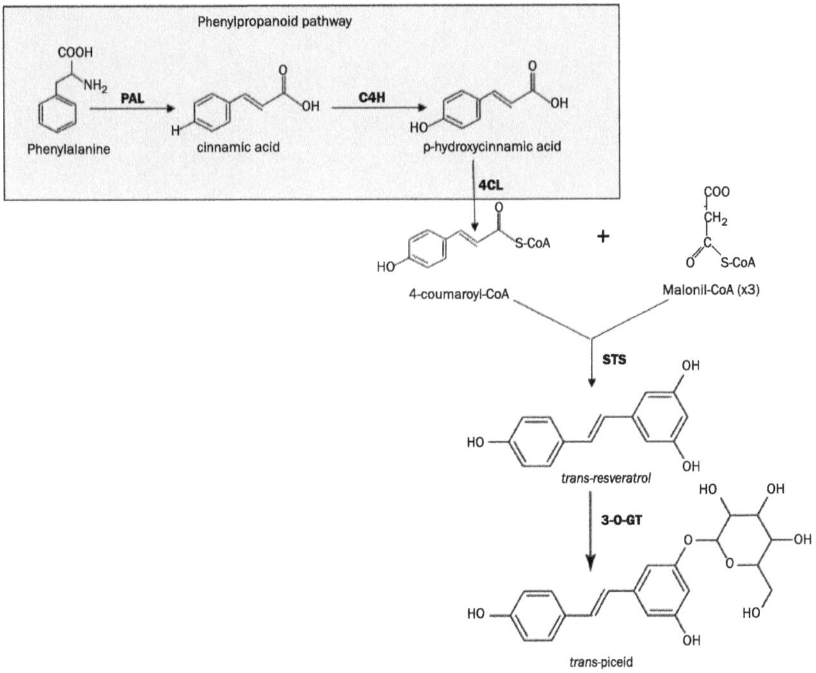

Figure 12.1 Resveratrol biosynthesis pathway

4-coumaroyl-CoA, in exceptional cases cinnamoyl-CoA), using subtly different reaction mechanisms. Based on the results of mutational and structural analysis and the fact that *STS* is found in a limited number of unrelated plant species, it has been suggested that *STS* has evolved from chalcone synthase (*CHS*) (Tropf et al., 1994). Indeed, CHS and STS share a deduced amino acid sequence similarity of up to 70 per cent (Schröder and Schröder, 1990; Eckermann et al., 2003).

Genes encoding *STS* have been isolated from grape (*Vitis vinifera* sp.) but are also found in peanut and pine. Members of the stilbene synthase (STS) family, responsible for the synthesis of resveratrol (Aggarwal et al., 2004), were previously described in the sequencing project (Jaillon et al., 2007). All 43 genes encoding *STSs* contained in the grapevine genome were profiled using the RNA-Seq method.

The stilbene *trans*-resveratrol (*trans*-3,5,4-trihydroxystilbene), a member of the stilbene family, has recently become the focus of a number of studies in medicine, plant physiology and plant biotechnology. Resveratrol exists in two stereoisoforms with *cis* or *trans* configuration, the latter being the most widely studied, although *cis*-resveratrol may also possess health-promoting properties. In nature, the most abundant form of resveratrol would appear to be 5,3,4′-dihydroxystilbene-3-O-β-D-glucopyranoside. The number as well as the position of moieties plays an important role in the biological activity of the compound (Soleas et al., 2001; Stivala et al., 2001; Frei, 2004). Originally, the stilbene type phyto-alexins such as resveratrol gained a lot of interest due to their fungicidal properties. In addition, resveratrol and its glycoside piceid are of great interest due to their health effects (Jang et al., 1997; Wang et al., 2006).

In plants, the synthesis of *trans*-resveratrol is induced by (a)biotic stresses, such as UV radiation, wounding or pathogen attack, and shows a remarkable antifungal activity (Sparvoli et al., 1994; Bais et al., 2000).

12.2 Health benefits of resveratrol

trans-Resveratrol has been associated with the 'French paradox' because its daily consumption, for example, in the form of red wine (Lippi et al., 2010a,b; Jandet et al., 2002; Stervbo et al. 2007), has been associated with beneficial effects (Wang et al., 2006) and protection against coronary diseases (Grønbæk et al., 2000). In addition, some plants have been modified genetically in order to produce *trans*-resveratrol for the assessment of its potential role in health promotion and plant disease control (Giovinazzo et al., 2012).

Resveratrol possesses numerous important bioactivities, including anti-inflammatory, antioxidant and anti-aggregatory functions, and modulation of lipoprotein metabolism (Frei, 2004). It has also been shown to possess chemopreventive properties against certain forms of cancer and cardiovascular disorders (Kundu and Surh, 2008). Subsequent work has shown that resveratrol extends the lifespans of lower eukaryotes (Gruber et al., 2007; Viswanathan et al., 2005; Wood et al., 2004). In mice, long-term administration of resveratrol induced gene expression patterns that resembled those induced by calorie restriction and delayed aging-related deterioration (Pearson et al., 2008). Resveratrol also decreased insulin resistance in type 2 diabetic patients (Brasnyó et al., 2011; Park et al., 2012), suggesting that the pathway targeted by resveratrol might be important for developing therapies for type 2 diabetes.

12.3 *trans*-resveratrol production through plant cell cultures

V. vinifera cell cultures have been used to increase *trans*-resveratrol production (Table 12.1). One of these strategies includes the use of biotic or abiotic elicitors such as UV light irradiation, β-cyclodextrins (CDs) and methyljasmonate (MJ). The effect of UV irradiation on stilbene content in grapevine suspension cultured cells has not yet been clarified, even though most of the research related to UV irradiation reported an increase in stilbene content in grape berries, leaves and callus tissue.

Jasmonic acid and its more active derivative MJ are signal molecules that act as key compounds of the signal transduction pathway of the plant defense mechanism. Thus, the production of secondary metabolites increases when plant cell cultures are elicited with jasmonates.

CDs are cyclic oligosaccharides that chemically resemble the alkyl-derived pectic oligosaccharides naturally released from the cell walls of fungal pathogens, and they act as true elicitors, since they activate some transcription factors in grapevine cells (Zamboni et al., 2009), inducing the production of *trans*-resveratrol. They have a hydrophilic external surface and a hydrophobic central cavity that traps *trans*-resveratrol, forming inclusion complexes. In addition, the high levels of *trans*-resveratrol accumulated in the culture medium have no toxic effect on the cell lines, allowing successful subcultures.

Significant progress in increasing *trans*-resveratrol content in plant cells has been reached using transformation of grape calli with the rolB gene of *A. rhizogenes* and the subsequent elicitation with CDs (Kiselev et al., 2007).

Taken together, these results indicate that efforts to obtain good yields of resveratrol should include a combination of different approaches and the identification of new, more effective and cheaper elicitors.

Furthermore, the production of secondary metabolites through plant cell cultures presents an additional challenge because the level of secondary metabolite production during long-term cultivation is often unstable and unpredictable. It has been demonstrated that cultivating plant cells in vitro for a long time induces various mutations, and genetic anomalies have been observed in both cultured cells and plants regenerated from cultured cells. For example, a three-year cultivation of rolB-transgenic grape cells resulted in a gradual reduction of production (Kiselev, 2011). Currently, the production of *trans*-resveratrol from cultivated cells ranges from 1 to 2% DW without and after elicitation, respectively.

Table 12.1 Bioproduction of resveratrol by plant cell cultures

Plant	Type of culture	Elicitors	Trans-resveratrol intracellular mg/L	Reference
Vitis vinifera L. cv Gamay	Cell suspension	DIMEB	100,3	Morales et al., 1998
		DIMEB + *X. ampelinus*	202,9	Morales et al., 1998
Vitis vinifera L. cv Barbera	Cell suspension	MeJA	0,11	Tassoni et al., 2005
Vitis vinifera L. cv Gamay rouge	Cell suspension	DIMEB	4,680	Bru and Pedreno 2006
		RAMEB	5,027	
Vitis vinifera L. cv Monastrell albino	Cell suspension	DIMEB	3,060	Bru and Pedreno 2006
		RAMEB	3,320	
Vitis Labrusca L. cv Concord	Cell suspension	L.Alanine	2,2	Chen et al., 2006
Vitis Amurensis Rupr.	Callus	Phenylalanine	2,4	Kiselev et al., 2007
		SA	4,7	
		MeJA	2,3	
		rolB gene	200–300	
Vitis vinifera L. cv Gamay Freaux	Cell suspension	MeJA+Sucrose	5,5	Belhadj et al., 2008
Vitis vinifera L. cv Monastrell albino	Cell suspension	DIMEB	753	Lijavetzky et al., 2008
		DIMEB+MeJA	3,651	
Vitis vinifera L. cv Barbera	Cell suspension	–	6,7	Ferri et al., 2009
Vitis amurensis Rupr.	Callus	ro1C gene	14,3	Dubrovina et al., 2010
Vitis amurensis Rupr.	Callus	5-azacytidine	5,1	Kiselev et al., 2010
Vitis vinifera L. cv Barbera	Cell suspension	sucrose	6,7	Ferri et al., 2011

12.4 Introducing new pathway branches to crop plants: *trans*-resveratrol synthesis in tomato fruits

STS structural genes have been transferred to a number of crops, to improve either the resistance to stresses or the nutritional value of the plant (for a review, see Giovinazzo et al., 2012 and Table 12.2).

Few articles have reported on the production of stilbenes in transgenic tomato, suggesting that it is indeed possible to introduce new branches of the flavonoid pathway, at least at its first step, by introducing foreign structural genes (Giovinazzo et al., 2005; Schijlen et al., 2006; Nicoletti et al., 2007). Transgenic tomato plants (*Solanum lycopersicum*, cv Moneymaker) expressing the stilbene synthase gene under constitutive 35S and mature fruit-specific promoter (TomLoxB) were obtained by genetic transformation through agrobacterium infection of cotyledons (Giovinazzo et al., 2005). The phenotype of all transformed lines was similar to that of the wild-type plants, showing regular development, flowering and fruit maturation. However, high resveratrol-producing, 35SS tomato fruits were seedless, whereas low-resveratrol LoxS fruit showed a normal seed set, comparable to wild type (D'Introno et al., 2009; Ingrosso et al., 2011). Free resveratrol and its glycosylated forms have both been detected in transgenic plants. Stilbene content also depends strongly upon plant species, probably on account of different endogenous pools of enzymes or precursors, as well as differences in secondary metabolic pathways. In resveratrol-synthesizing tomato lines, the free to glycosylated resveratrol ratio was related to the fruit ripening stage (Giovinazzo et al., 2005). Furthermore, these related compounds accumulate differentially in different fruit tissues at the mature stage (Nicoletti et al., 2007).

In transgenic tomato, resveratrol synthesis was able to increase the overall antioxidant properties of the fruit, as well as the ascorbate/glutathione content, with a consequent two–threefold increase in antioxidant activity of fruits, and a correlation was found between resveratrol concentrations and antioxidant capacities in the ripening stages, accumulating high resveratrol levels (Giovinazzo et al., 2005).

The effects of resveratrol-enriched tomato extracts on cyclooxygenase (COX)-2 expression induced by phorbol ester in monocyte–macrophage U937 cells indicated that resveratrol reduces the level of the inducible, but not the constitutive, COX isoform, thus confirming and expanding the anti-inflammatory activity of resveratrol. It is noteworthy that

Table 12.2 Bioproduction of resveratrol by plant metabolic engineering

Plant species	Gene	µg/g	Biological activity	Reference
Solanum lycopersicum L.	StSy	53 (in fruit tissues)	Increased antioxidant capability	Giovinazzo et al., 2005
	STS	30 (in fruit tissues)	Food quality improvement	Schijlen et al., 2006
	StSy	50–120 (in fruit tissues)	Modulation of other polyphenols	Nicoletti et al., 2007
	StSy	10–120 (in fruit tissues) 50–180 (in flower tissues)	Increased antioxidant and anti-inflammatory capabilities and male sterility	D'Introno et al., 2009
Brassica napus L.	Vst1	361–616	Food quality improvement	Hüsken et al., 2005
Humulus lupulus L.	Vst1	490–560	Modulation of other polyphenols	Schwekendiek et al., 2007
Rehmannia glutinosa Libosch	AhRS3	Up to 650 (under stress)	Antioxidant capability	Lim et al., 2005

Published by Woodhead Publishing Limited, 2013

transgenic tomato fruits, with different levels of resveratrol, showed higher antioxidant and anti-inflammatory properties than wild-type fruits (D'Introno et al., 2009)

12.5 Concluding remarks

trans-Resveratrol bioproduction has attracted considerable attention, with regard to potential applications for human benefit, and to better understanding the molecular mechanisms of *trans*-resveratrol biosynthesis for their potential exploitation. The *trans*-resveratrol metabolic pathway has now been well characterized, and the expression and activities of the genes involved have been shown to be enhanced in response to elicitation (Lijavetzky et al., 2008). Molecular engineering of *trans*-resveratrol production in plants using *STS* genes has been achieved recently, and this approach constitutes a valuable strategy for increasing not only the resistance of plants to disease, but also the nutritional value of agricultural crops and food products.

Biotechnology using tissues or plant cells thus represents environmentally and ecologically friendly strategies, thereby allowing easier and safer purification, as well as pesticide-free production methods (Delaunois et al., 2009; Halls and Yu, 2008). The advantages afforded by in vitro plant cell culture systems result from their putative low production costs, considering that the only equipment plant cells need is a stirred bioreactor.

Moreover, secondary metabolites can be produced constitutively, or induced by elicitation. Resveratrol itself has been produced in large amounts from grape cell cultures, and this has been carried out using elicitation with cyclodextrins, which also act as stabilizers, thereby compensating for the slow growth of plant cells (Lucas-Abellan et al., 2007).

Nevertheless, the bioproduction of *trans*-resveratrol and other stilbene derivatives remains a considerable challenge. In the future, scale-up to industrial processes has to be achieved and optimized, in order to develop a sustainable production of *trans*-resveratrol and other stilbenes of interest.

Major crop plants eaten for nutrition often lack the desired polyphenols or contain only small amounts in the relevant tissues. Metabolic engineering has provided a means to improve flavonoid composition as well as content. The quality of crop plants, nutritionally or otherwise, is a direct function of this metabolite content. Research to improve the nutritional quality of plants has historically been limited by a lack of basic knowledge of plant metabolism and the compounding challenge of resolving the complex interactions of thousands of metabolic pathways. However, despite all the extensive work, there are still challenging tasks ahead (Martin et al., 2012). The isolation and cloning of most of the structural or regulatory genes open up possibilities to develop plants with tailor-made optimized metabolite levels and composition.

12.6 References

Aggarwal BB, Bhardwaj A, Aggarwal RS, Seeram NP, Shishodia S, et al. (2004) Role of resveratrol in prevention and therapy of cancer: preclinical and clinical studies. *Anticancer Res.* 24:2783–840.

Bais AJ, Murphy PJ, Dry IB (2000) The molecular regulation of stilbene phytoalexin biosynthesis in *Vitis vinifera* during grape berry development. *Austral. J. Plant Physiol.* 27:425–33.

Belhadj A, Telef N, Saigne C, Cluzet S, Barrieu F, et al. (2008) Effect of methyl jasmonate in combination with carbohydrates on gene expression of PR proteins, stilbene and anthocyanin accumulation in grapevine cell cultures. *Plant Physiol. Biochem.* 46:493–9.

Brasnyó P, Molnár GA, Mohás M, Markó L, Laczy B, et al. (2011) Resveratrol improves insulin sensitivity, reduces oxidative stress and activates the Akt pathway in type 2 diabetic patients. *Br. J. Nutr.* 106:383–9.

Bru MR, Pedreno GMLDE (2006) Method for the production of resveratrol in cell cultures. US 2006/0205049 A1.

Burns J, Yokota T, Ashihara H, Lean ME, Crozier A (2002) Plant foods and herbal sources of resveratrol. *J. Agric. Food Chem.* 50:3337–40.

Chen JX, Hall DE, Murata J, De Luca V (2006) L-Alanine induces programmed cell death in *V. labrusca* cell suspension cultures. *Plant Sci.* 171:734–44.

Constant J (1997) Alcohol, ischemic heart disease, and the French paradox. *Coron. Artery Dis.* 8:645–9.

Counet C, Callemien D, Collin S (2006) Chocolate and cocoa: New sources of *trans*-resveratrol and *trans*-piceid. *Food Chem.* 98:649–57.

Delaunois B, Conreux A, Clément C, Jeandet P (2009) Molecular engineering of resveratrol in plants. *Plant Biotechnol. J.* 7:2–12.

D'Introno A, Paradiso A, Scoditti E, D'Amico L, De Paolis A, et al. (2009) Antioxidant and anti-inflammatory properties of tomato fruit synthesizing different amount of stilbenes. *Plant Biotechnol. J.* 7:422–9.

Dubrovina AS, Manyakhin AY, Zhuravlev YN, Kiselev KV (2010) Resveratrol content and expression of phenylalanine ammonialyase and stilbene synthase genes in rolC transgenic cell cultures of *Vitis amurensis*. *Appl. Microbiol. Biotechnol.* 88:727–36.

Eckermann C, Matthes B, Nimtz M, Reiser V, Lederer B, et al. (2003) Covalent binding of chloroacetamide herbicides to the active site cysteine of plant type III polyketide synthases. *Phytochemistry* 64(6):1045–54.

Ferri M, Tassoni A, Franceschetti M, Righetti L, Naldrett MJ, et al. (2009) Chitosan treatment induces changes of protein expression profile and stilbene distribution in *Vitis vinifera* cell suspensions. *Proteomics* 9:610–24.

Ferri M, Righetti L, Tassoni A (2011) Increasing sucrose concentrations promote phenylpropanoid biosynthesis in grapevine cell cultures. *J. Plant Physiol.* 168:189–95.

Frei B (2004) Efficacy of dietary antioxidants to prevent oxidative damage and inhibit chronic disease. *J. Nutr.* 134:3196–8.

Giovinazzo G, Ingrosso I, Paradiso A, De Gara L, Santino A (2012) Resveratrol biosynthesis: plant metabolic engineering for nutritional improvement of food. *Plant Foods Hum. Nutr.* doi: 10.1007/s11130-012-0299-8 [Epub ahead of print].

Giovinazzo G, D'Amico L, Paradiso A, Bollini R, Sparvoli F, et al. (2005) Antioxidant metabolite profiles in tomato fruit constitutively expressing the grapevine stilbene synthase gene. *Plant Biotechnol. J.* 3:57–69.

Grønbæk M, Becker U, Johansen D, Gottschau A, Schnohr P, et al. (2000) Type of alcohol consumed and mortality from all causes, coronary heart disease and cancer. *Ann. Intern. Med.* 133:411–19.

Gruber J, Tang SY, Halliwell B (2007) Evidence for a trade-off between survival and fitness caused by resveratrol treatment of *Caenorhabditis elegans*. *Ann. N.Y. Acad. Sci.* 1100:530–42.

Halls C, Yu O (2008) Potential for metabolic engineering of resveratrol biosynthesis. *Trends Biotechnol.* 26:77–81.

Hertog MGL, Hollman PC, Katan MB, Kromhout D (1993) Intake of potentially anticarcinogenic flavonoids and their determinants in adults in the Netherlands. *Nutr. Cancer* 20:21–9.

Hüsken A, Baumert A, Milkowski C, Becker HC, Strack D, et al. (2005) Resveratrol glucoside (Piceid) synthesis in seeds of transgenic oilseed rape (Brassica napus L.). *Theor. Appl. Genet.* 111:1553–62.

Ingrosso I, Bonsegna S, De domenico S, Laddomada B, Blando F, et al. (2011) Novel findings into parthenocarpy in tomato: a stilbene synthase approach to induce male sterility. *Plant Physiol. Biochem.* 49:1092–9.

Jaillon O, Aury JM, Noel B, Policriti A, Clepet C, et al. (2007) The grapevine genome sequence suggests ancestral hexaploidization in major angiosperm phyla. *Nature* 449:463–7.

Jandet P, Douillet-Breuil AC, Bessis R, Debord S, Sbaghi M, et al. (2002) Phytoalexins from the vitaceae: Biosynthesis, phytoalexin gene expression in transgenic plants, antifungal activity, and metabolism. *J Agric. Food Chem.* 50:2731–41.

Jang M, Cai L, Udeani GO, Slowing KV, Thomas CF, et al. (1997) Cancer chemopreventive activity of resveratrol, a natural product derived from grapes. *Science* 275:218–20.

Kiselev KV (2011) Perspectives for production and application of resveratrol. *Appl. Microbiol. Biotechnol.* 90:417–25.

Kiselev KV, Dubrovina AS, Veselova MV, Bulgakov VP, Fedoreyev SA, et al. (2007) The rolB gene-induced overproduction of resveratrol in *Vitis amurensis* transformed cells. *J. Biotechnol.* 128:681–92.

Kiselev KV, Tyunin AP, Manyakhin AY, Zhuravlev YN (2011) Resveratrol content and expression patterns of stilbene synthase genes in *Vitis amurensis* cells treated with 5-azacytidine. *Plant Cell Tissue Organ Cult.* 105:65–72.

Koes RE, Quattrocchio F, Mol JNM (1994) The flavonoid biosynthetic pathway in plants: function and evolution. *BioEssays* 16:123–32.

Kundu JK, Surh YJ (2008) Cancer chemopreventive and therapeutic potential of resveratrol: mechanistic perspectives. *Cancer Lett.* 269:243–61.

Lippi G, Franchini M, Favaloro EJ, Targher G (2010a) Moderate red wine consumption and cardiovascular disease risk: beyond the "French paradox". *Semin. Thromb. Hemost.* 36:59–70.

Lippi G, Franchini M, Guidi GC (2010b) Red wine and cardiovascular health: The "French Paradox" revisited. *Int. J. Wine Res.* 2:1–7.

Lijavetzky D, Almagro L, Belchi-Navarro S, Martinez-Zapater JM, Bru R, et al. (2008) Synergistic effect of methyljasmonate and cyclodextrin on stilbene biosynthesis pathway gene expression and resveratrol production in Monastrell grapevine cell cultures. *BMC Res. Notes* 1:132.

Lucas-Abellan C, Fortea I, López-Nicolás JM, Núñez-Delicado E (2007) Cyclodextrins as resveratrol carrier system. *Food Chem.* 104:39–44.

Lim JD, Yun SJ, Chung IM, Yu CY (2005) Resveratrol synthase transgene expression and accumulation of resveratrol glycoside in *Rehmannia glutinosa*. *Mol. Breed.* 16:219–33.

Martin C, Butelli E, Petroni K, Tonelli C (2012) How can research on plants contribute to promoting human health? *The Plant Cell* 23:1685–99.

Morales M, Bru R, Garcia-Carmona F, Barcelo AR, Pedreno MA (1998) Effect of dimethyl-β-cyclodextrins on resveratrol metabolism in Gamay grapevine cell cultures before and after inoculation with Xylophilus ampelinus. *Plant Cell Tissue Org. Cult.* 53:179–87.

Muir SR, Collins GJ, Robinson S, Hughes S, Bovy A, et al. (2001) Over-expression of petunia chalcone isomerase in tomato results in fruit containing increased levels of flavonols. *Nature Biotechnol.* 19:470–4.

Nicoletti I, DeRossi A, Giovinazzo G, Corradini D (2007) Identification and quantification of stilbenes in fruits of transgenic tomato plants (Lycopersicon esculentum Mill.) by reversed phase HPLC with photodiode array and mass detection. *J. Agric. Food Chem.* 55:3304–11.

Park SJ, Ahmad F, Philp A, Baar K, Williams T, et al. (2012) Resveratrol ameliorates aging-related metabolic phenotypes by inhibiting cAMP phosphodiesterases. *Cell* 14:421–33.

Pearson KJ, Baur JA, Lewis KN, Peshkin L, Price NL, et al. (2008) Resveratrol delays age-related deterioration and mimics transcriptional aspects of dietary restriction without extending life span. *Cell Metab.* 8:157–68.

Schijlen EGWM, de Vos CHR, Jonker H, van den Broeck H, Molthoff J, et al. (2006) Pathway engineering for healthy phytochemicals leading to the production of novel flavonoids in tomato fruit. *Plant Biotechnol. J.* 4:433–44.

Schröder J and Schröder G (1990) Stilbene and chalcone synthases: related enzymes with key functions in plant-specific pathways. *Z Naturforsch.* 45:1–8.

Schwekendiek A, Spring O, Heyerick A, Pickel B, Pitsch NT, et al. (2007) Constitutive expression of a grapevine stilbene synthase gene in transgenic hop (*Humulus lupulus* L.) yields resveratrol and its derivatives in substantial quantities. *J. Agric. Food Chem.* 55:7002–9.

Soleas GJ, Diamandis EP, Goldberg DM (2001) The world of resveratrol. *Adv. Exp. Med. Biol.* 492:159–82.

Sparvoli F, Martin C, Scienza A, Gavazzi G, Tonelli C (1994) Cloning and molecular analysis of structural genes involved in flavonoid and stilbene biosynthesis in grape (*Vitis vinifera* L.). *Plant Mol. Biol.* 24:743–55.

Stervbo U, Vang O, Bonnesen C (2007) A review of the content of the putative chemopreventive phytoalexine resveratrol in red wine. *Food Chem.* 101:449–57.

Stivala LA, Savio M, Carafoli F, Perucca P, Bianchi L, et al. (2001) Specific structural determinants are responsible for the antioxidant activity and the cell cycle effects of resveratrol. *J. Biol. Chem.* 276:22 586–94.

Tassoni A, Fornale S, Franceschetti M, Musiani F, Michael AJ, et al. (2005) Jasmonates and Na-orthovanadate promote resveratrol production in *Vitis vinifera* cv. Barbera cell cultures. *New Phytol.* 166:895–905.

Tropf S, Lanz T, Rensing SA, Schröder J, Schröder G (1994) Evidence that stilbene synthases have developed from chalcone synthases several times in the course of evolution. *J. Mol. Evol.* 38(6):610–18.

Viswanathan M, Kim SK, Berdichevsky A, Guarente L (2005) A role for SIR-2.1 regulation of ER stress response genes in determining *C. elegans* life span. *Dev. Cell* 9:605–15.

Wang J, Ho L, Zhao Z, Seror I, Humala N, et al. (2006) Moderate consumption of Cabernet Sauvignon attenuates Abeta neuropathology in a mouse model of Alzheimer's disease. *FASEB J.* 20:2313–20.

Wood JG, Rogina B, Lavu S, Howitz K, Helfand SL, et al. (2004) Sirtuin activators mimic caloric restriction and delay ageing in metazoans. *Nature* 430:686–9.

Zamboni A, Gatto P, Cestaro A, Pilati S, Viola R, et al. (2009) Grapevine cell early activation of specific responses to DIMEB, a resveratrol elicitor. *BMC Genomics* 10:363.

Index

abiotic stress, 42–3
Agilent SurePrint 3G, 171
Agrobacterium tumefaciens, 131
agronomic traits, 55
angiosperms, 69–70
Arabidopsis, 131
artificial microRNA (amiRNA), 126
Auxin Responsive Elements (ARE), 176

β-cyclodextrins, 226
bacterial artificial chromosome (BAC), 73
basic-helix-loop-helix 1 (CIB1), 102
biodiversity, 205–6
bioproduction, 223–31
biotechnology, 121, 206–7
biotic stress, 42–3
Brassica napus, 194–5
   cold tolerance selection, 196–7
   double haploidy utilisation in resistance selection, 188–91
   haploid technology breeding application, 183–97
   improved seed meal selection, 193–6
   in vitro mutagenesis, 186–7
   isolated microspore culture technique, 184–5
   method, 185–6
   modified seed oil composition selection, 191–3

breeding
   cold tolerance selection, 196–7
   double haploidy utilisation in resistance selection, 188–91
   doubled haploid method for *Brassica napus*, 185–6
   doubled haploid technology application of *Brassica napus*, 183–97
   improved seed meal selection, 193–6
   in vitro mutagenesis, 186–7
   isolated microspore culture technique, 184–5
   modified seed oil composition selection, 191–3
bud organogenesis, 155–6

cabbage seedpod weevil, 188–9
canola oil, 191–2
canola seed, 194
cap analysis of gene expression (CAGE), 6–7
carbon assimilation, 208
cell-to-cell communication, 174
central oscillator, 99
cereals
   heat stress tolerance, 207–10
      elevated temperature effect on plants, 207–10
   modern approaches for yield and food security under temperature stress, 210–15

Index

molecular markers and breeding, 210–12
Ceutorhynchus obstrictus, 188–9
chalcone synthase, 223–4
chi-square goodness-of-fit test, 195
chilling injury, 169
ChillPeach, 171
chloroplast DNA (cpDNA), 75–6
chromosome doubling, 185
circadian and clock associated 1 (CCA1), 99
cis-resveratrol, 225
Clavata 3, 174
climacteric fruits, 168, 169
clonal selection, 121
coat protein-mediated resistance (CP-MR), 125
cold tolerance, 196–7
competing endogenous RNA (ceRNA), 26
conifer leaves, 70–1
conifer tracheids, 71
conifers, 67–86
   chemical divergence, 82–5
   functional differentiation, 69–71
     biological difference with angiosperms, 70
     genome function, 76–82
     genome structure and composition, 72–6
     genome size, 72
   systems biology approach, 85–6
Conserved Ortholog Sets (COS), 75
conventional breeding, 123–4
copy DNA (cDNA), 3
copy number variation (CNV), 159–60
CoT analysis, 73
crop improvement
   metabolomics, 55–9
   microRNA role, 25–6

cross-breeding, 121
CRY1 gene, 98, 100, 103, 108
CRY1a gene, 98, 109–10
CRY1b gene, 98
CRY2 gene, 98, 100, 102–3, 107–10
CRY2 transgenic over-expressor (CRY2-OX), 107–10
CRY3 gene, 98
cryptochrome
   diurnal global transcription profiles in tomato, 105–10
   functions, 99–100
   light-regulated gene expression in plants, 100–5
   tomato transcriptome modulation, 97–110
CTG134, 175–6
cultivars, 120–1
cup-shaped cotyledon 2 (CUC2), 22
cyclic oligosaccharides, 226

DCL1, 18, 20
DeepSuperSAGE, 5, 6
diurnal global transcription profiles, 105–10
DNA coding sequences, 4
DNA microarray, 4
DNA typing, 155–6
domestication, 151–2
double stranded RNA (dsRNA), 128
doubled haploid (DH)
   application in *Brassica napus* breeding, 183–97
   breeding method for *Brassica napus*, 185–6
   cold tolerance selection, 196–7
   improved seed meal selection, 193–6
   in vitro mutagenesis, 186–7
   isolated microspore culture technique, 184–5

modified seed oil composition
  selection, 191–3
  utilisation in resistance selection, 188–91
drupe
  peach ripening transcriptomics for quality improvement, 165–78

early flowering 4 (ELF4), 106
embryonic flower2 (EMF2), 34
ENOD40, 4
epigenetics, 31–43
expressed sequence tag (EST), 4, 211, 212–13
expression Quantitative Trait Loci (eQTL), 7
Extra sex combs (Esc), 34

fertilisation independent endosperm (FIE), 34
fertilisation independent SEED2 (FIS2), 34
flavonols, 85
flowering locus C, 40
flowering locus T, 102
Food and Agriculture Organisation (FAO), 187
French paradox, 225
functional genomics, 7–10, 212–13
  application of systems biology and omics technology in plant crops, 7

gas phase mass spectrometry (GC/MS), 50
gene expression, 78–80
  light-regulated, 100–5
gene expression profiling, 79
gene redundancy, 159–60
genetic diversity, 206
genetic engineering, 128, 214–15
genetic heterogeneity, 153
genetic improvement technology, 124
genetic mapping, 75
genetic transformation technology, 135
genetically modified plants (GMP), 134
genome sequencing, 85–6, 153–4, 154–5, 172–3
genome wide association mapping, 122
genome wide association studies (GWAS), 57–8
genomics
  grapevine, 119–37, 149–60
  selection, 122
germplasm, 122, 194
gibberellic acid (GA), 104
global climate, 207–8
GOLVEN peptides *see* Auxin Responsive Elements (ARE)
Grape ReSeq Consortium, 155
grapevine, 119–37, 149–60
  bud organogenesis, somatic mutations and DNA typing of somatic chimeras, 155–6
  copy number variation, gene redundancy and subtle specialisation, 159–60
  genetic and molecular markers for genetic diversity and genome selection studies, 120–2
  genome analysis, 154–5
  genomic tools in genome sequencing era, 153–4
  inserted transgenes stability, 135–7
  large structural variation using NGS, 158–9
  origin of *Vitis vinifera*, domestication and early selection for fruit characters, 151–2
  phenotypic variation sources, 153

phenotypically divergent clones and underlying DNA variation, 156–8
transgene insertion loci identification and characterisation, 131–3
transgene silencing, 128–31
transposon insertion-site profiling using NGS, 158
vector backbone integration, 133–5
grapevine breeding, 123–8
expression cassettes of plant transformation vectors, 127
Grapevine fanleaf virus (GFLV), 125–6

haplotype, 152
heat shock proteins (HSP), 208
heat stress tolerance
cereals, 207–10
elevated temperature effect on plants, 207–10
heat stress transcription factors (HSF), 209
hermaphroditism, 152
high performance liquid chromatography (HPLC), 188–9
hormone–photoreceptor crosstalk, 109
HSP101, 209
HT-SuperSAGE, 5
HY5 homologue (HYH), 102
Hypocotyl in Far-Red 1 (HFR1), 102

in vitro embryo germination, 185
in vitro embryogenesis, 185
in vitro mutagenesis, 186–7
Institute of Applied Genomics, 155
integrated pest management (IPM), 189
International Atomic Energy Agency (IAEA) Mutant Varieties Database, 187
International Maize Wheat Improvement Centre, 211–12
isolated microspore culture technique, 184–5

jasmonic acid, 226

large structural variation, 158–9
late elongated hypocotyl (LHY), 106
lignin, 84
like heterochromatin protein1 (LHP1), 36
linkage disequilibrium (LD), 151–2
liquid phase mass spectrometry for negative ionisation (LC/MS-NEG), 52
liquid phase mass spectrometry for positive ionisation (LC/MS-POS), 52
liquid phase mass spectrometry (LC/MS), 50
long hypocotyl 5 (HY5), 102
long hypocotyl (LHY), 99
long terminal repeat (LTR), 74
LTR retrotransposons, 74

macro-co-linearity, 75
macro-synteny, 75
marker-assisted breeding, 211
marker assisted selection, 122
mass spectrometry (MS), 50
metabolite profiling, 49–60
metabolomic platform, 52–4
metabolomics in crop improvement, 55–9
metabolomics in plant science, 54–5
methodological approach, 50–2
metabolomics, 52–4, 213–14

Index

crop improvement, 55–9
   metabolite association with genotype, 57–9
   metabolite biomarker identification, 59
   nitrogen use efficiency, 56–7
   plant science, 54–5
methylation–filtration technique, 73
1-methylcyclopropene (1-MCP), 169–70
methyljasmonate (MJ), 226
microarray hybridisation, 171
microarray-transcriptomic profiling, 105, 170–4
   chromosomal location of ethylene-related genes on peach physical map, 173
   expression profiles of ethylene-related genes during fruit development and ripening, 172
microRNA (miRNA), 8–9, 80–1, 128
   functions, 20–5
      stress response and development, 21
   gene evolution, 19
   gene transcription, 16
   modes of action, 18–19
   plant *vs.* animal, 20
   plants, 15–26
   processing, 17–18
      biogenesis in plant cells, 17
   roles in crop improvement, 25–6
microsatellite loci *see* simple sequence repeat (SSR)
microsatellites, 157
microspore
   culture technique, 184–5
   extraction, 185
minor allele frequency (MAF), 155
modified seed oil composition, 191–3
molecular characterisation, 126–7

mosaicism *see* somatic chimerism
μPEACH1.0, 171
μPEACH3.0, 171

Nepovirus, 125
next-generation sequencing (NGS), 122, 151, 158–9
nitrogen use efficiency (NUE), 56–7
non-climacteric fruits, 168
non-coding RNA (ncRNA), 4
   environmental stress response, 40–1
   Polycomb complex regulatory cofactor, 38–40
      acting in *cis* and *trans* to recruit PRC2, 39
nuclear cap-binding complex (CBC), 18
nuclear magnetic resonance (NMR), 50

omics technology, 3–10
   advancement, 7–10
   cap analysis of gene expression (CAGE), 6–7
   SuperSAGE, 5–6
open reading frames (ORF), 24

P-starvation-responsive (PSR) genes, 23
pathogen derived resistance (PDR), 125–6
pathogen-mediated resistance, 124–5
peach
   development and ripening, 169–70
   fruit, 167–8
   Microarray Transcript Profiling, 170–4
   ripening control, 174–8
   ripening transcriptomics for improvement of drupe quality, 165–78
Pepper mild mottle virus, 125

pest resistance, 188–91
  cabbage seedpod weevil, 188–9
  root maggot, 189
  Sclerotinia sclerotiorum, 190–1
PHABULOSA (PHB), 22
phenotypic variation, 153
  grapevine bud sports, varieties and wild relatives, 149–60
photoexcitation, 101
photomorphogenesis, 101
photoreceptors, 99
photosynthesis, 208–9
phytochromes, 98, 105
plant biodiversity, 205–15
  approaches in cereals for yield and food security under temperature stress, 210–15
  future trends, 215
  heat stress tolerance in cereals, 207–10
plant biotechnology, 205–15
  approaches in cereals for yield and food security under temperature stress, 210–15
  future trends, 215
  heat stress tolerance in cereals, 207–10
plant genomics, 3–10
  advancement in functional genomics, 7–10
  cap analysis of gene expression (CAGE), 6–7
  SuperSAGE, 5–6
plant research
  metabolite profiling, 49–60
  methodological approach, 50–2
plants
  circadian clock, 99
  microRNA, 15–26
  Polycomb proteins epigenetics control, 31–43

pleiohomeotic repressive complex (Pho-RC), 32
Polycomb group (PcG) proteins
  abiotic and biotic stress response, 42–3
  epigenetics control, 31–43
  function in three dimensional nuclear organisation, 41–2
  multi-protein complexes, 32–6
    core components of the PRC complexes, 33
    vernalisation process and the PRC2 complexes, 35
  non-coding RNA (ncRNA), 38–40, 40–1
  plant development, 36–7
Polycomb repressive complex 1 (PRC1), 32–3, 35–6
Polycomb repressive complex 2 (PRC2), 32–4
polymerase chain reaction (PCR), 126, 207
post-transcriptional gene silencing (PTGS), 129
post-transcriptional regulation, 80
pre-transcriptional regulation, 80
primary miRNA (pri-miRNA), 16, 17–18
priming, 43
protein thermo-stability, 208–9
proteomics, 56, 213–14
protoplast electroporation, 126
Prunus persica, 166, 169
pseudo response regulator 7 (PRR7), 106
pseudo response regulator (PRR) protein, 99

Quality Protein Maize (QPM), 211
quantitative trait loci (QTL), 7, 57, 121, 189, 211

Index

random amplified polymorphic DNA marker (RAPD), 192
rapeseed oil, 191–2
reference transcriptome, 77
reiterated reproductive meristem (RRM), 157
resins, 82–3
restriction fragment length polymorphism (RFLP), 192
resveratrol
   biosynthesis pathways, 224
   health benefits, 225
   natural bioproduction, 223–31
   stilbenes, 223–5
   trans-resveratrol production through plant cell cultures, 226–7
   trans-resveratrol synthesis in tomato fruits, 228–30
RGL1, 176–8
Riboreg, 4
RNA-directed DNA methylation (RdDM), 129
RNA-induced silencing complex (RISC), 18, 129
RNA-induced transcriptional gene silencing (RITS) complex, 129
RNA interference, 8–9
RNA sequencing (RNAseq), 171
RNA silencing, 8–9, 129
root growth factor-like peptide *see* RGL1
root growth factor (RGF), 175
root maggot, 189

Sclerotinia sclerotiorum, 190–1
SE (SERRATE) protein, 18
secondary metabolites, 82, 84
SEPALLATA-SPL genes, 21
serial analysis of gene expression (SAGE), 5–6

short-interfering RNA (siRNA), 80–1
short vegetative phase (SVP), 36
signal transduction, 101–2
simple sequence repeat (SSR), 121
sinapine, 196
single nucleotide mutation, 157
single nucleotide polymorphism (SNP), 58, 122, 154, 192
small heat-shock proteins (sHSP), 78
small interfering RNA (siRNA), 8–9, 128
small RNA (sRNA), 80–2
Solanum lycopersicum see tomato
somatic chimerism, 156
somatic mutations, 155–6
species diversity, 206
split-read method, 158
stilbenes, 84, 223–5
superoxide dismutase 2 (CSD2), 23
SuperSAGE, 5–6
   protocol scheme, 5
suppressor of phytochrome A (SPA1), 106

tannins, 84–5
targeting induced local lesions in genomes (TILLING), 8
TERMINAL FLOWER 1 (TFL1), 157
terpenoids, 83
terrestrial vascular plants, 167
timing of CAB1 (TOC1), 99
Tobacco mosaic virus (TMV), 125
tobamoviruses, 125
tomato, 228
   cryptochrome gene modulation of global transcriptome, 97–110
   diurnal global transcription profiles, 105–10
Tomato Systemin (SYS), 174
TOMATOMA, 8
TomLoxB, 228

torus-margo pit membrane, 71
tracheary element, 174
trans-3,5,4-trihydroxystilbene *see trans*-resveratrol
trans-resveratrol, 224–5, 225, 230
   production through plant cell cultures, 226–7
     bioproduction, 227
     synthesis in tomato fruits, 228–30
     bioproduction by plant metabolic engineering, 229
transacting siRNA (tasiRNA), 19
transcription
   microRNA (miRNA) gene, 16
transcription factors, 104
transcriptional gene silencing (TGS), 129
transcriptome
   tomato cryptochrome genes modulation, 97–110
transcriptomics, 56, 76–8
   peach ripening for improvement of drupe quality, 165–78
transferred DNA (T-DNA), 126, 128, 131–3, 133–5
transgene expression, 136
transgene insertion loci, 131–3
transgene silencing, 128–31

transgenic tomato, 228
transposable elements (TE), 158
transposon insertion-site profiling, 158

ultra high performance liquid chromatography (UHPLC), 52
untranslated region (UTR), 20
UV irradiation, 226
UV radiation, 187

vector backbone sequences (VBS), 133–5
vegetative propagation, 152
vernalisation 2 (VRN2), 34
Vitis Microsatellite Consortium (VMC), 121
Vitis vinifera, 150, 224, 226
   origin, domestication and early selection for fruit characters, 151–2
volatile-rich resins, 82–3

xylem, 70–1, 78
   transcriptome, 78

yellow-seeded *Brassica napus*, 194–6
yellow-seeded rapeseed, 194–6

CPSIA information can be obtained at www.ICGtesting.com
Printed in the USA
BVOW10*0851070414

349827BV00001B/1/P